Other Monographs in this Series

No. 1. Irving Kaplansky: *Algebraic and analytic aspects of operator algebras*
2. Gilbert Baumslag: *Lecture notes on nilpotent groups*
3. Lawrence Markus: *Lectures in differentiable dynamics*
4. H. S. M. Coxeter: *Twisted honeycombs*
5. George W. Whitehead: *Recent advances in homotopy theory*
6. Walter Rudin: *Lectures on the edge-of-the-wedge theorem*
7. Yozo Matsushima: *Holomorphic vector fields on compact Kähler manifolds*
8. Peter Hilton: *Lectures in homological algebra*
9. I. N. Herstein: *Notes from a ring theory conference*
10. Branko Grünbaum: *Arrangements and spreads*
11. Irving Glicksberg: *Recent results on function algebras*
12. Barbara L. Osofsky: *Homological dimensions of modules*
13. Michael Rabin: *Automata on infinite objects and Church's problem*
14. Sigurdur Helgason: *Analysis on Lie groups and homogeneous spaces*
15. R. G. Douglas: *Banach algebra techniques in the theory of Toeplitz operators*
16. Joseph L. Taylor: *Measure algebras*
17. Louis Nirenberg: *Lectures on linear partial differential equations*
18. Avner Friedman: *Differential games*

Conference Board of the Mathematical Sciences
REGIONAL CONFERENCES SERIES IN MATHEMATICS

supported by the
National Science Foundation

Number 19

UNITARY DILATIONS OF HILBERT SPACE OPERATORS AND RELATED TOPICS

by

BÉLA SZ.-NAGY

Published for the
Conference Board of the Mathematical Sciences
by the
American Mathematical Society
Providence, Rhode Island

Expository Lectures
from the CBMS Regional Conference
held at the University of New Hampshire
June 7–11, 1971

AMS (MOS) Subject Classifications (1970). Primary 47A20, 47A15, 47A45, 47A60, 47A65; Secondary 30A76, 30A78, 46E20.

Library of Congress Cataloging in Publication Data CIP

Szőkefalvi-Nagy, Béla, 1913–
Unitary dilations of Hilbert space operators and related topics.
(Regional conferences series in mathematics, no. 19)
"Expository lectures from the CBMS regional conference held at the University of New Hampshire, June 7–11, 1971."
Bibliography: p.
1. Linear operators. 2. Hilbert space. 3. Functions, Characteristic. I. Title. II. Series.
QA1.R33 no. 19 [QA329.2] 515'.73 73–17332
ISBN 0–8218–1669–1

Printed in the United States of America

CONTENTS

CONTENTS

Preface

This paper organizes and expounds the lectures which the author delivered at the Regional Conference held 7–11 June 1971 at the University of New Hampshire under the sponsorship of the Conference Board of the Mathematical Sciences and with the support of the National Science Foundation.

The subject is taken from the theory of unitary dilations of contraction operators and of the theory of functional models of such operators. The fundamental result in this area, namely the existence and uniqueness of the minimal unitary dilation for a contraction, was proved by the author in 1953 and was the starting point of various far-reaching investigations of the structure and properties of Hilbert space operators. This trend is far from being exhausted. Most of the investigations in this direction were made by the author in close and lasting collaboration with Ciprian Foiaş; our book on *Harmonic analysis of operators on Hilbert space,** quoted in the sequel by [H], gives a detailed account of our results as well as of a good part of the relevant results of a number of other mathematicians.

In my lectures I selected some of the chapters of the theory which I expounded in some detail, other chapters were only sketched, and many others not even mentioned. I have tried to be quite detailed in the geometric (or purely operator theoretic) aspects of the theory, thus preparing a natural and rapid introduction, by "Fourier representations," of the characteristic functions and functional models of operators. I have tried to make clear the connections of the invariant subspace problem with the factorization problem of the characteristic function, but I could not enter into such applications as the fairly complete spectral theory of weak contractions. I have included the elements of the functional calculus of contractions and of the theory of operators of C_0, and have indicated the main results in the Jordan model theory of these operators, but I have not included one-parameter continuous semigroups, or the way of passing from contractions to (not necessarily bounded) dissipative or accretive operators. I have given some attention to the "lifting theorem," which is a most useful tool in many applications, but I have not even mentioned applications of the theory to scattering

---

* North-Holland and Akadémiai Kiadó, 1970. Revised and enlarged edition of the first edition, in French, published by Masson et Akadémiai Kiadó in 1967. A Russian edition of the revised variant was published in 1970 by "Mir" in Moscow.

theory or to prediction theory, etc. I have indicated the problem and the most interesting results concerning unitary dilations of a commutative system of contractions but have excluded any reference to dilations of representations of function algebras, or to unitary ρ-dilations.

Thus, in short, I did not endeavour to give a bird's eye view of *all* the realm of operator theory connected with the dilation concept; instead I tried to give a possibly clear picture of some of its main provinces. I hope that some of the readers will find the theme interesting and promising enough for further study, with the help of the book [H] and the current literature.

Szeged
July 1973

Béla Sz.-Nagy

1. Isometric and Unitary Dilations of a Contraction Operator

1. In the first two sections the Hilbert spaces which we consider are either all real or all complex; they may be either separable or not. Operators are always linear and bounded.

Let A and B be operators on the Hilbert spaces $\mathfrak{A}$ and $\mathfrak{B}$, respectively. We call B a *dilation* of A (also, a "strong" or "power" dilation) if $\mathfrak{A}$ is a subspace of $\mathfrak{B}$ and the following condition is satisfied:

$$A^n = P_{\mathfrak{A}} B^n | \mathfrak{A} \qquad (n = 0, 1, \cdots), \tag{1.1.1}$$

$P_{\mathfrak{A}}$ denoting orthogonal projection from $\mathfrak{B}$ onto $\mathfrak{A}$. Condition (1.1.1) is equivalent to the condition

$$(A^n h_1, h_2) = (B^n h_1, h_2) \qquad (h_1, h_2 \in \mathfrak{A};\ n = 0, 1, 2, \cdots); \tag{1.1.2}$$

hence if B is a dilation of A, then B^* is a dilation of A^*.

Condition (1.1.1) is satisfied in particular if

$$AP_{\mathfrak{A}} = P_{\mathfrak{A}} B \tag{1.1.3}$$

holds (on $\mathfrak{B}$); indeed, (1.1.3) implies $A^n P_{\mathfrak{A}} = P_{\mathfrak{A}} B^n$ for $n = 0, 1, \cdots$, and restriction of both sides to $\mathfrak{A}$ yields (1.1.1). Conversely, (1.1.1) implies

$$AP_{\mathfrak{A}}(B^n a) = AA^n a = A^{n+1} a = P_{\mathfrak{A}} B^{n+1} a = P_{\mathfrak{A}} B(B^n a)$$

for $a \in \mathfrak{A}$ and $n = 0, 1, \cdots$; hence (1.1.3) holds when both sides are restricted to the subspace $\bigvee_{n=0}^{\infty} B^n \mathfrak{A}$ (of $\mathfrak{B}$). Thus if this subspace equals $\mathfrak{B}$ (i.e., if $\mathfrak{B}$ has no proper subspace containing $\mathfrak{A}$ and invariant for $\mathfrak{B}$), then conditions (1.1.1) and (1.1.3) are *equivalent.*

The importance of the above notion comes from the fact that an operator A of general type may have a dilation B of some quite special type, and hence the study of A can be reduced to the study of this B of special type.

2. For a contraction operator T on the Hilbert space $\mathfrak{H}$ (i.e., with $\|T\| \leq 1$) we define the *defect operator D, defect space $\mathfrak{D}$ and defect index $\mathfrak{d}$* by

$$D = (I - T^*T)^{1/2}, \qquad \mathfrak{D} = \overline{D\mathfrak{H}}, \qquad \mathfrak{d} = \dim \mathfrak{D},$$

the bar denoting closure. Let D_*, $\mathfrak{D}_*$, $\mathfrak{d}_*$ have the corresponding meaning for T^*. The obvious equation $T(I - T^*T) = T - TT^*T = (I - TT^*)T$ implies $Tp(D^2) = p(D_*^2)T$ for any polynomial $p(\lambda)$, and hence for any continuous function on $0 \leq \lambda \leq 1$, in particular for the function $\lambda^{1/2}$. This gives the equations

$$TD = D_* T, \qquad DT^* = T^* D_*; \tag{1.2.1}$$

whence we deduce

$$T\mathfrak{D} \subset \mathfrak{D}_*, \qquad T^*\mathfrak{D}_* \subset \mathfrak{D}. \tag{1.2.2}$$

Next we prove

Theorem 1. *Every contraction operator T on a Hilbert space $\mathfrak{H}$ has a unitary dilation U on a Hilbert space $\mathfrak{K}$. One can require that this unitary dilation be minimal in the sense*

$$\bigvee_{n=-\infty}^{\infty} U^n\mathfrak{H} = \mathfrak{K}; \tag{1.2.3}$$

U is then determined by T uniquely (that is, up to an isometric isomorphism leaving the vectors of $\mathfrak{H}$ invariant).

Proof. Form the Hilbert space $\mathfrak{K}$ of vectors

$$k = \left\langle \cdots, h_{-2}, h_{-1}, \boxed{h_0}, h_1, h_2, \cdots \right\rangle \tag{1.2.4}$$

with components $h_0 \in \mathfrak{H}$, $h_n \in \mathfrak{D}$, $h_{-n} \in \mathfrak{D}_*$ for $n \geq 1$, and norm

$$\|k\| = \left(\sum_{-\infty}^{\infty} \|h_n\|^2 \right)^{1/2} < \infty,$$

and embed $\mathfrak{H}$ in $\mathfrak{K}$ by identifying $h \in \mathfrak{H}$ with $\left\langle \cdots, 0, \boxed{h}, 0, \cdots \right\rangle$. The orthogonal projection $P_{\mathfrak{H}}$ is then given by $P_{\mathfrak{H}} k = \left\langle \cdots, 0, \boxed{h_0}, 0, \cdots \right\rangle = h_0$. Define on $\mathfrak{K}$ the operators U, U' by

$$Uk = \left\langle \cdots, h_{-3}, h_{-2}, \boxed{D_* h_{-1} + Th_0}, -T^* h_{-1} + Dh_0, h_1, h_2, \cdots \right\rangle \tag{1.2.5}$$

and

$$U'k = \left\langle \cdots, h_{-1}, D_* h_0 - Th_1, \boxed{T^* h_0 + Dh_1}, h_2, h_3, \cdots \right\rangle; \tag{1.2.6}$$

these definitions are correct because of (1.2.2). Immediate computations based on the relations (1.2.1) yield that U is isometric and $UU' = I_{\mathfrak{K}}$; hence we conclude that U is *unitary* and $U' = U^{-1} = U^*$.

From (1.2.5) it readily follows that

$$U^n h = \left\langle \cdots, 0, \boxed{T^n h}, DT^{n-1} h, \cdots, Dh, 0, \cdots \right\rangle$$

and hence

$$T^n h = P_{\mathfrak{H}} U^n h \quad \text{for } h \in \mathfrak{H} \text{ and } n = 0, 1, \cdots.$$

Thus U is a unitary dilation of T.

Next observe that, for any $h \in \mathfrak{H}$,

$$\begin{aligned} Uh - Th &= U\langle \cdots, 0, \boxed{h}, 0, \cdots \rangle - \langle \cdots, 0, \boxed{Th}, 0, \cdots \rangle \\ &= \langle \cdots, 0, \boxed{Th}, Dh, 0, \cdots \rangle - \langle \cdots, 0, \boxed{Th}, 0, \cdots \rangle \\ &= \langle \cdots, 0, \boxed{0}, Dh, \cdots \rangle \end{aligned}$$

and, by analogous reasoning,

$$U^* h - T^* h = \langle \cdots, 0, D_* h, \boxed{0}, 0, \cdots \rangle;$$

therefore, the subspaces

$$(1.2.7) \qquad \mathfrak{L} = \overline{(U - T)\mathfrak{H}} \quad \text{and} \quad \mathfrak{L}^* = \overline{(U^* - T^*)\mathfrak{H}}\,[1]$$

consist respectively of the vectors $\langle \cdots, 0, \boxed{0}, d, 0, \cdots \rangle$ with arbitrary $d \in \mathfrak{D}$, and of the vectors $\langle \cdots, 0, d_*, \boxed{0}, 0, \cdots \rangle$ with arbitrary $d_* \in \mathfrak{D}_*$. It then follows that, for $n \geq 0$,

$$U^n \mathfrak{L} = \left\{ \langle \cdots, 0, \boxed{0}, \overset{1}{\widetilde{0}}, \cdots, 0, \overset{n+1}{\widetilde{d}}, 0, \cdots \rangle : d \in \mathfrak{D} \right\} \quad \text{and}$$

$$U^{*n} \mathfrak{L}^* = \left\{ \langle \cdots, 0, \overset{-n-1}{\widetilde{d_*}}, 0, \cdots, \overset{-1}{\widetilde{0}}, \boxed{0}, 0, \cdots \rangle : d_* \in \mathfrak{D}_* \right\}.$$

Bearing also in mind the way $\mathfrak{H}$ was embedded in $\mathfrak{K}$ we conclude that $\mathfrak{K}$ has the following decomposition:

$$(1.2.8) \qquad \mathfrak{K} = \cdots \oplus U^{-2}\mathfrak{L}^* \oplus U^{-1}\mathfrak{L}^* \oplus \mathfrak{L}^* \oplus \mathfrak{H} \oplus \mathfrak{L} \oplus U\mathfrak{L} \oplus U^2\mathfrak{L} \oplus \cdots.$$

From (1.2.7) and (1.2.8) it follows readily that the minimality condition (1.2.3) is also fulfilled.

It remains to prove uniqueness. Let U' and U'' be any two minimal unitary dilations of T, say on $\mathfrak{K}'$ and $\mathfrak{K}''$, and observe first that

$$\begin{aligned} (U'^n h_1, U'^m h_2) &= (U'^{n-m} h_1, h_2) = (T^{n-m} h_1, h_2) \\ &= (U''^{n-m} h_1, h_2) = (U''^n h_1, U''^m h_2) \end{aligned}$$

for any $h_1, h_2 \in \mathfrak{H}$ and for $n \geq m$; equality of the two extreme members holds— by the

---

[1] Note that in this notation the star in $\mathfrak{L}^*$ does not mean adjugation: it is used only to indicate symmetry in the definitions of $\mathfrak{L}$ and $\mathfrak{L}^*$.

the symmetry property of the inner product—in case $n \leq m$ also. Hence we conclude that the map

$$\sum_{-\infty}^{\infty} U'^n h_n \mapsto \sum_{-\infty}^{\infty} U''^n h_n \qquad (h_n \in \mathfrak{H};\ h_n = 0 \text{ for } |n| \text{ large enough})$$

is isometric, and therefore extends by continuity to an isometry ϕ of $\mathfrak{K}'$ onto $\mathfrak{K}''$. This isometry leaves the vectors in $\mathfrak{H}$ invariant, and carries U' into U'', i.e., $\phi U' = U''\phi$. Thus if we disregard such isometric isomorphisms ϕ, the minimal unitary dilation of T is unique.

In the sequel one may, but need not, restrict oneself to any special realization of the minimal unitary dilation U of T. The subspaces $\mathfrak{L}$ and $\mathfrak{L}^*$ as defined by (1.2.7) are *wandering* [2] for U and the decomposition (1.2.8) holds, independently of the special realization of $\mathfrak{K}$ and U.

3. The following equality is of importance:

$$\mathfrak{H} \oplus \mathfrak{L} = U\mathfrak{H} \oplus U\mathfrak{L}^*. \tag{1.3.1}$$

It suffices to prove that

$$\mathfrak{H} \oplus (U - T)\mathfrak{H} = U\mathfrak{H} \oplus (I - UT^*)\mathfrak{H}. \tag{1.3.2}$$

The orthogonality of $U\mathfrak{H}$ and $U\mathfrak{L}^*$ follows from the orthogonality of $\mathfrak{H}$ and $\mathfrak{L}^*$ (see (1.2.8)), and equality (1.3.2) follows from the fact that every vector k of $\mathfrak{K}$ which can be written in one of the forms

$$k = h' + (U - T)h'' \qquad (h', h'' \in \mathfrak{H}),$$

$$k = Uh_1 + (I - UT^*)h_2 \qquad (h_1, h_2 \in \mathfrak{H})$$

can be written in the other also: one has only to set

$$h_1 = T^*h' + (I - T^*T)h'', \qquad h' = Th_1 + (I - TT^*)h_2,$$
$$\text{or}$$
$$h_2 = h' - Th'', \qquad h'' = h_1 - T^*h_2.$$

4. One of the consequences of (1.3.1) is that $U\mathfrak{H} \subset \mathfrak{H} \oplus \mathfrak{L}$, whence $U^n\mathfrak{H} \subset \mathfrak{H} \oplus \mathfrak{L} \oplus \cdots \oplus U^{n-1}\mathfrak{L}$ $(n = 0, 1, \cdots)$. On the other hand, we clearly have $\mathfrak{L} \subset \mathfrak{H} \vee U\mathfrak{H}$, whence $U^n\mathfrak{L} \subset U^n\mathfrak{H} \vee U^{n+1}\mathfrak{H}$ $(n = 0, 1, \cdots)$. We conclude that the subspaces $\mathfrak{H} \oplus M_+(U; \mathfrak{L})$ and $\bigvee_0^\infty U^n\mathfrak{H}$ of $\mathfrak{K}$ are equal. Denote this subspace by $\mathfrak{K}_+$ and set $U_+ = U|\mathfrak{K}_+$. Clearly U_+ is an *isometric* dilation of T, indeed a *minimal* one, that is, satisfying

---

[2] A subspace $\mathfrak{A}$ of the space of an isometry V is called "wandering" if $\mathfrak{A}, V\mathfrak{A}, V^2\mathfrak{A}, \cdots$ are mutually orthogonal. Then we denote the orthogonal sum $\bigoplus_0^\infty V^n\mathfrak{A}$ by $M_+(V; \mathfrak{A})$; if there is no ambiguity, we also write $M_+(\mathfrak{A})$. In case V is unitary, $V = U$, we write $M(U; \mathfrak{A})$ or $M(\mathfrak{A})$ for $\bigoplus_{-\infty}^\infty U^n\mathfrak{A}$.

(1.4.1)
$$\mathfrak{K}_+ = \bigvee_0^\infty U_+^n \mathfrak{H}.$$

Therefore, recalling (1.1.3) we have

(1.4.2)
$$TP_+ = P_+ U_+$$

where P_+ denotes the orthogonal projection of $\mathfrak{K}_+$ onto $\mathfrak{H}$.

One shows in analogy to the unitary case that isometric dilations of T satisfying this kind of minimality condition are isometrically isomorphic so that the minimal isometric dilation of T is (essentially) *unique.*

Note that if we start with the representation of $\mathfrak{K}$ and U as given by formulas (1.2.4) and (1.2.5) then $\mathfrak{K}_+$ will consist of the vectors $k \in \mathfrak{K}$ with zero components h_n $(n \le -1)$ so that, by a natural embedding, $\mathfrak{K}_+$ can be considered as the space of vectors

$$k = \langle h_0, h_1, h_2, \cdots \rangle \quad \text{with } h_0 \in \mathfrak{H},\ h_n \in \mathfrak{D} \quad (n \ge 1),$$

and $\|k\| = (\Sigma_0^\infty \|h_n\|^2)^{1/2}$, and U_+ is then defined by

(1.4.3)
$$U_+ k = \langle Th_0, Dh_0, h_1, h_2, \cdots \rangle.$$

It is an easy exercise to verify directly that this operator is a minimal isometric dilation of T.

In the sequel we shall not keep to a particular realization of the minimal isometric dilation U_+ of T, but we shall always consider it as the restriction of the minimal unitary dilation U. We shall have

(1.4.4)
$$\mathfrak{K}_+ = \mathfrak{H} \oplus M_+(\mathfrak{L}).$$

Let us now consider the Wold decomposition of $\mathfrak{K}_+$ corresponding to the isometry U_+, i.e., the decomposition

(1.4.5)
$$\mathfrak{K}_+ = M_+(\mathfrak{L}_*) \oplus \mathfrak{R},$$

where

(1.4.6)
$$\mathfrak{L}_* = \mathfrak{K}_+ \ominus U_+\mathfrak{K}_+ \quad \text{and } \mathfrak{R} = \bigcap_0^\infty U_+^n \mathfrak{K}_+.$$

The Wold decomposition (1.4.4) *reduces* the operator U_+; U_+ is a unilateral shift on $M_+(\mathfrak{L})$ and unitary on $\mathfrak{R}$. For the wandering subspace $\mathfrak{L}_*$ generating the unilateral shift part of U_+ we obtain, using the equalities $\mathfrak{K}_+ = \mathfrak{H} \oplus M_+(\mathfrak{L}) = \mathfrak{H} \oplus \mathfrak{L} \oplus U_+M_+(\mathfrak{L})$ and $U_+\mathfrak{K}_+ = U_+(\mathfrak{H} \oplus M_+(\mathfrak{L})) = U_+\mathfrak{H} \oplus U_+M_+(\mathfrak{L})$, that

$$\mathfrak{L}_* = (\mathfrak{H} \oplus \mathfrak{L}) \ominus U_+\mathfrak{H}.$$

Because of (1.3.1) we therefore have

$$\mathfrak{L}_* = \overline{(I - UT^*)\mathfrak{H}}. \tag{1.4.7}$$

As $M_+(\mathfrak{L})$ is invariant for U_+, its orthogonal complement in $\mathfrak{K}_+$, that is $\mathfrak{H}$, is invariant for U_+^*. As by the dilation property we have $P_{\mathfrak{H}} U_+^* | \mathfrak{H} = T^*$, we conclude that

$$T^* = U_+^* | \mathfrak{H}. \tag{1.4.8}$$

We close this section by observing that because of the relations

$$\mathfrak{K} = \bigvee_{-\infty}^{\infty} U^n \mathfrak{H} = \bigvee_{-\infty}^{-1} U^n \mathfrak{K}_+$$

we obtain, from (1.4.5), the following alternative decomposition of $\mathfrak{K}$:

$$\mathfrak{K} = M(\mathfrak{L}_*) \oplus \mathfrak{R}. \tag{1.4.9}$$

We shall use the above decompositions of the spaces $\mathfrak{K}_+$ and $\mathfrak{K}$, and the relation (1.4.8) between T and U_+ to construct a functional model for T.

2. Further Properties of the Minimal Unitary Dilation

1. For a contraction operator T on $\mathfrak{H}$ and for any $h \in \mathfrak{H}$ the sequence $\|T^n h\|$ $(n = 0, 1, \cdots)$ is nonincreasing and therefore convergent. If U is a unitary dilation of T (on $\mathfrak{K}$) then we have, for $n \geq m \longrightarrow \infty$,

$$\|U^{-m}T^m h - U^{-n}T^n h\|^2 = \|T^m h\|^2 - 2\,\mathrm{Re}\,(U^{n-m}T^m h,\ T^n h) + \|T^n h\|^2$$
$$= \|T^m h\|^2 - \|T^n h\|^2 \longrightarrow 0.$$

Hence, the strong limit

(2.1.1) $$L = \lim_{n \to \infty} U^{-n}T^n$$

exists and is an operator from $\mathfrak{H}$ into $\mathfrak{K}$.

We claim that L *is the orthogonal projection from* $\mathfrak{H}$ *to the orthogonal complement of* $M(\mathfrak{L})$ *in* $\mathfrak{K}$.

To this end we first observe that if $h, h' \in \mathfrak{H}$ and $n \geq 0$, $n \geq \nu$, then

$$(U^{-n}T^n h,\ U^{-\nu}(U - T)h') = (T^n h,\ U^{n-\nu+1}h' - U^{n-\nu}Th')$$
$$= (T^n h,\ T^{n-\nu+1}h' - T^{n-\nu}Th') = 0,$$

and hence $Lh \perp U^{-\nu}\mathfrak{L}$ for all ν and therefore $Lh \perp M(\mathfrak{L})$. On the other hand, we have

$$h - U^{-n}T^n h = \sum_{m=0}^{n-1} (U^{-m}T^m h - U^{-m-1}T^{m+1}h) = \sum_{m=0}^{n-1} U^{-m-1}(U - T)T^m h$$

and hence (as $n \longrightarrow \infty$)

(2.1.2) $$h - Lh = \sum_{m=0}^{\infty} U^{-m-1}(U - T)T^m h \quad (\in M(\mathfrak{L})).$$

This concludes the proof of our assertion.

By analogous reasoning,

(2.1.3) $$L_* = \lim_{n \to \infty} U^n T^{*n}$$

exists and is the orthogonal projection from $\mathfrak{H}$ to $M(\mathfrak{L}_*)^\perp$, i.e., by virtue of (1.4.9), to $\mathfrak{R}$. Analogous to (2.1.2) we have

$$h - L_* h = \sum_{m=0}^{\infty} U^m (I - UT^*) T^{*m} h. \tag{2.1.4}$$

From the definitions (2.1.1) and (2.1.3) of L and L_* and from their projection properties we easily deduce

$$\begin{aligned} T^n h \to 0\ (n \to \infty) &\Leftrightarrow Lh = 0 \Leftrightarrow h \in M(\mathfrak{L}), \\ T^{*n} h \to 0\ (n \to \infty) &\Leftrightarrow L_* h = 0 \Leftrightarrow h \in M(\mathfrak{L}_*). \end{aligned} \tag{2.1.5}$$

We introduce classes of contraction operators T on $\mathfrak{H}$ according to their asymptotic behaviour as $n \to \infty$:

$$C_{0\cdot} = \{T:\ T^n h \to 0 \text{ for } all\ h \in \mathfrak{H}\}, \qquad C_{\cdot 0} = \{T:\ T^{*n} h \to 0 \text{ for } all\ h \in \mathfrak{H}\},$$

$$C_{1\cdot} = \{T:\ T^n h \to 0 \text{ for } h = 0\ only\}, \qquad C_{\cdot 1} = \{T:\ T^{*n} h \to 0 \text{ for } h = 0\ only\};$$

furthermore, set

$$C_{\alpha\beta} = C_{\alpha\cdot} \cap C_{\cdot\beta}.$$

Then, by (2.1.5),

$$T \in C_{0\cdot} \Leftrightarrow \mathfrak{H} \subset M(\mathfrak{L}),$$

$$T \in C_{\cdot 0} \Leftrightarrow \mathfrak{H} \subset M(\mathfrak{L}_*).$$

2. If U is *minimal* then, obviously, $\mathfrak{H} \subset M(\mathfrak{L}) \Leftrightarrow \mathfrak{R} = M(\mathfrak{L})$, and $\mathfrak{H} \subset M(\mathfrak{L}_*) \Leftrightarrow \mathfrak{R} = M(\mathfrak{L}_*)$ so that we conclude in this case that

$$\begin{aligned} \mathfrak{R} = M(\mathfrak{L}) &\Leftrightarrow T \in C_{0\cdot}, \\ \mathfrak{R} = M(\mathfrak{L}_*) &\Leftrightarrow T \in C_{\cdot 0}. \end{aligned} \tag{2.2.1}$$

Still in the case of the minimal unitary dilation U of T consider the span $M(\mathfrak{L}) \vee M(\mathfrak{L}_*)$ or rather its orthogonal complement $\mathfrak{H}_0 = M(\mathfrak{L})^\perp \cap M(\mathfrak{L}_*)^\perp$. As $\mathfrak{H}_0$ is in particular orthogonal to $U^n \mathfrak{L}$ and $U^{-n} \mathfrak{L}^*$ $(n = 0, 1, \cdots)$ it follows from the decomposition (1.2.8) of $\mathfrak{R}$ that $\mathfrak{H}_0 \subset \mathfrak{H}$. Now a vector $h \in \mathfrak{H}$ belongs to $\mathfrak{H}_0$ if and only if it is orthogonal to $U^{-n}(U - T)\mathfrak{H}$ and $U^n (U^* - T^*)\mathfrak{H}$ for $n = 1, 2, \cdots$, i.e., to $(U^{*m} - U^{*m+1}T)\mathfrak{H}$ and $(U^m - U^{m+1}T)\mathfrak{H}$ for $m = 0, 1, \cdots$; using the dilation property this is equivalent to the conditions $\|T^m h\| = \|T^{m+1} h\|$ and $\|T^{*m} h\| = \|T^{*m+1} h\|$ $(m = 0, 1, \cdots)$. Hence, $\mathfrak{H}_0$ is composed of the vectors h of $\mathfrak{H}$ for which the following equations hold:

$$\cdots = \|T^{*2} h\| = \|T^* h\| = \|h\| = \|Th\| = \|T^2 h\| = \cdots. \tag{2.2.2}$$

Now conditions (2.2.2) characterize the maximal subspace of $\mathfrak{H}$ which reduces the contraction operator T to a unitary operator (theorem of Langer and Foiaş). If this subspace is trivial (i.e., equals $\{0\}$) then T is called *completely nonunitary* (cnu). Hence, *if T is cnu then* $\mathfrak{H}_0 = \{0\}$ and therefore

(2.2.3) $$\mathfrak{K} = M(\mathfrak{L}) \vee M(\mathfrak{L}_*).$$

3. Suppose in the sequel that T is *completely nonunitary* (cnu) so that (2.2.3) holds.

Also, recall equation (1.4.9) telling that $\mathfrak{K} = M(\mathfrak{L}_*) \oplus \mathfrak{R}$. Using (2.2.3) we infer

$$\mathfrak{R} = P_{\mathfrak{R}}\mathfrak{K} = P_{\mathfrak{R}}[M(\mathfrak{L}) \vee M(\mathfrak{L}_*)] = \overline{P_{\mathfrak{R}}M(\mathfrak{L})}$$

because $\mathfrak{R} \perp M(\mathfrak{L}_*)$. Let us set

(2.3.1) $$Q = P_{M(\mathfrak{L}_*)}|M(\mathfrak{L}),$$

i.e., Q is the orthogonal projection from $M(\mathfrak{L})$ into $M(\mathfrak{L}_*)$. Then,

(2.3.2) $$\mathfrak{R} = \overline{(I - Q)M(\mathfrak{L})}.$$

Q intertwines the parts of U in $M(\mathfrak{L})$ and in $M(\mathfrak{L}_*)$, i.e.,

(2.3.3) $$Q(U|M(\mathfrak{L})) = (U|M(\mathfrak{L}_*))Q.$$

This follows from the fact that the decomposition $\mathfrak{K} = M(\mathfrak{L}_*) \oplus \mathfrak{R}$ reduces U. For, if $l = l_* + r$ is the corresponding decomposition of an $l \in M(\mathfrak{L})$, then clearly $Ul = Ul_* + Ur$ will be the decomposition of Ul, and this means that $QUl = UQl$.

Next we remark that

(2.3.4) $$QM_+(\mathfrak{L}) \subset M_+(\mathfrak{L}_*).$$

Indeed, one infers from the decomposition (1.2.8) of $\mathfrak{K}$ that $M_+(\mathfrak{L})$ is orthogonal to $U^{-n}\mathfrak{L}^*$ $(n = 0, 1, \cdots)$, i.e., to $U^{-m}\mathfrak{L}_*$ $(m = 1, 2, \cdots)$.

It is of importance to compute Ql in particular for l of the form $l = (U - T)h$ $(h \in \mathfrak{H})$, since these vectors are dense in $\mathfrak{L}$, and since, as a consequence of (2.3.3), the values of Q on $\mathfrak{L}$ determine its values on the whole $M(\mathfrak{L})$.

Now, as a result of (2.1.4) we have

(2.3.5) $$h = \sum_0^\infty U^m(I - UT^*)T^{*m}h + L_*h \quad \text{for every } h \in \mathfrak{H}.$$

Apply U termwise to (2.3.5) to obtain an expansion of Uh, and give in (2.3.5) the role of h to Th to obtain an expansion of Th. So we get

$$Uh = \sum_0^\infty U^{m+1}(I - UT^*)T^{*m}h + UL_*h,$$

$$Th = \sum_0^\infty U^m(I - UT^*)T^{*m}Th + L_*Th,$$

and hence

$$(2.3.6) \qquad Q(U - T)h = -(I - UT^*)Th + \sum_1^\infty U^m(I - UT^*)T^{*m-1}D^2h,$$

because $UL_*h - L_*Th \in \mathfrak{R}$ (recall that L is the projection from $\mathfrak{H}$ into $\mathfrak{R}$, and $\mathfrak{R}$ is invariant for U).

Formula (2.3.6) will be useful in the sequel.

In connection with the operator Q let us observe some more facts:

The first is an immediate consequence of the definition of Q as a projection: Q is an isometry iff $M(\mathfrak{L}) \subset M(\mathfrak{L}_*)$. In case T is cnu, this condition is equivalent (by virtue of (2.2.3)) to the condition that $\mathfrak{R} = M(\mathfrak{L}_*)$, and hence, by virtue of (2.2.1), to the condition that $T \in C_{\cdot 0}$. Thus, for a cnu T, we have

$$(2.3.7) \qquad Q \text{ is an isometry} \Leftrightarrow T \in C_{\cdot 0}.$$

The second observation is about the orthogonal complement of $QM_+(\mathfrak{L})$ in $M_+(\mathfrak{L}_*)$. Let $l_* \in M_+(\mathfrak{L}_*)$. Then,

$$l_* \perp QM_+(\mathfrak{L}) \Leftrightarrow l_* \; (= Ql_*) \perp M_+(\mathfrak{L}) \Leftrightarrow l_* \in \mathfrak{H}$$

(cf. (1.4.4)). Now for an $h \in \mathfrak{H}$ we have $h \in M_+(\mathfrak{L}_*)$ iff $h \in M(\mathfrak{L}_*)$ (by virtue of (1.2.8)), and hence iff $T^{*n}h \to 0$ (cf. (2.1.5)). So we conclude

$$(2.3.8) \qquad M_+(\mathfrak{L}_*) \ominus \overline{QM_+(\mathfrak{L})} = \{h: \; h \in \mathfrak{H}, \; T^{*n}h \to 0 \; (n \to \infty)\}.$$

In particular we have

$$(2.3.9) \qquad \overline{QM_+(\mathfrak{L})} = M_+(\mathfrak{L}_*) \Leftrightarrow T \in C_{\cdot 1}.$$

3. Characteristic Function and Functional Model

1. Recall some of the results of §§1 and 2 for a contraction operator T on $\mathfrak{H}$, its minimal unitary dilation U on $\mathfrak{K}$, and minimal isometric dilation U_+ on $\mathfrak{K}_+$.

(α) $\mathfrak{L} = \overline{(U-T)\mathfrak{H}}$ and $\mathfrak{L}_* = \overline{(I-UT^*)\mathfrak{H}}$ are wandering subspaces for U as well as for U_+.

(β) $\mathfrak{K}_+ = M_+(\mathfrak{L}_*) \oplus \mathfrak{R}$ and $\mathfrak{K} = M(\mathfrak{L}_*) \oplus \mathfrak{R}$, where $\mathfrak{R}$ is the space of the unitary part of U_+; cf. (1.4.5) and (1.4.9).

(γ) $\mathfrak{H} = \mathfrak{K}_+ \ominus M_+(\mathfrak{L})$ and $T^* = U_+^*|\mathfrak{H}$; cf. (1.4.4) and (1.4.8).

Moreover, the operator

$$Q = P_{M(\mathfrak{L}_*)}|M(\mathfrak{L}) \quad \text{(orthogonal projection } M(\mathfrak{L}) \to M(\mathfrak{L}_*))$$

satisfies the following:

(δ) $UQl = QUl$ for $l \in M(\mathfrak{L})$.

(ϵ) $QM_+(\mathfrak{L}) \subset M_+(\mathfrak{L}_*)$; cf. (2.3.4).

(η) $Q(U-T)h = -(I-UT^*)Th + \Sigma_1^\infty U^m(I-UT^*)T^{*m+1}D_T^2 h$ ($h \in \mathfrak{H}$); cf. (2.3.6).

(ι) (for cnu T) Q is an isometry iff $T \in C_{\cdot 0}$; cf. (2.3.7).

(κ) $\overline{QM_+(\mathfrak{L})} = M_+(\mathfrak{L}_*)$ iff $T \in C_{\cdot 1}$; cf. (2.3.9).

Finally (again for cnu T),

(μ) $\mathfrak{R} = \overline{(I-Q)M(\mathfrak{L})}$; cf. (2.3.2).

2. By using appropriate Fourier representations we translate these relations to obtain a function space model for T, at least if T is *completely nonunitary* on a *separable* space $\mathfrak{H}$. All spaces to be considered from now on will be *complex*.

First observe that the dilation property of U implies

$$\|(U-T)h\| = \|Dh\| \quad \text{and} \quad \|(I-UT^*)h\| = \|D_*h\| \qquad (h \in \mathfrak{H}). \tag{3.2.1}$$

These relations imply that the transformations

$$(U-T)h \mapsto Dh, \quad (I-UT^*)h \mapsto D_*h \qquad (h \in \mathfrak{H}) \tag{3.2.2}$$

can be extended by continuity to *unitary* maps

$$\phi\colon \mathfrak{L} \to \mathfrak{D}, \qquad \phi_*\colon \mathfrak{L}_* \to \mathfrak{D}_*. \tag{3.2.3}$$

Now for any separable Hilbert space $\mathfrak{A}$ denote by $L^2(\mathfrak{A})$ the space of $\mathfrak{A}$-vector valued, measurable and norm-square integrable functions on the unit circle (with normalized Lebesgue measure). Let $H^2(\mathfrak{A})$ be the Hardy subspace of $L^2(\mathfrak{A})$. Denote by χ the function

$$\chi(e^{it}) \equiv e^{it}.$$

We infer from (3.2.3) that the transformations

$$(3.2.4) \qquad \Phi\colon \sum_m U^m l_m \mapsto \sum_m \chi^m(\phi l_m), \qquad \Phi_*\colon \sum_m U^m l_{*m} \mapsto \sum_m \chi^m(\phi_* l_{*m}),$$

where $l_m \in \mathfrak{L}$, $l_{*m} \in \mathfrak{L}_*$, $\Sigma \|l_m\|^2 < \infty$, $\Sigma \|l_{*m}\|^2 < \infty$, are unitaries,

$$(3.2.5) \qquad \Phi\colon M(\mathfrak{L}) \to L^2(\mathfrak{D}), \qquad \Phi_*\colon M(\mathfrak{L}_*) \to L^2(\mathfrak{D}_*).$$

Observe that Φ maps $M_+(\mathfrak{L})$ onto $H^2(\mathfrak{D})$, and Φ_* maps $M_+(\mathfrak{L}_*)$ onto $H^2(\mathfrak{D}_*)$. Moreover we obviously have

$$(3.2.6) \qquad \Phi U|M(\mathfrak{L}) = \chi \cdot \Phi, \qquad \Phi_* U|M(\mathfrak{L}_*) = \chi \cdot \Phi_*.$$

Consider the operator

$$\Theta = \Phi_* Q \Phi^{-1};$$

this is a contraction of $L^2(\mathfrak{D})$ into $L^2(\mathfrak{D}_*)$ which, as a result of property (ϵ) of Q, maps $H^2(\mathfrak{D})$ into $H^2(\mathfrak{D}_*)$. By virtue of property (δ) of Q and by (3.2.4), Θ commutes with multiplication by χ (and, as a consequence, with multiplication by any scalar valued bounded measurable function).

From these properties of the operator Θ, it is possible to deduce (by Lebesgue integral theory methods) that Θ itself is a multiplication operator, i.e.,

$$(\Theta u)(e^{it}) = \Theta(e^{it})u(e^{it}) \qquad (u \in L^2(\mathfrak{D})),$$

where $\Theta(e^{it})$ is the radial limit (in the strong topology for operators), almost everywhere on the unit circle, of a contractive analytic function

$$\Theta(\lambda) = \Theta_0 + \lambda\Theta_1 + \lambda^2\Theta_2 + \cdots \qquad (|\lambda| < 1),$$

whose values are operators $\mathfrak{D} \to \mathfrak{D}_*$.

We can obtain the coefficients Θ_k by using (η) and the definitions (3.2.2)–(3.2.5). Indeed, if $h \in \mathfrak{H}$, then we have

$$\Theta D h = \Phi_* \underset{\sim}{Q} \Phi^{-1} D h = \Phi_* \underset{\sim}{Q}(U - T)h$$

$$= \Phi_* \left[-(I - UT^*)Th + \sum_1^\infty U^m(I - UT^*)T^{*m-1}D^2 h \right]$$

$$= -D_* Th + \sum_1^\infty \chi^m D_* T^{*m-1} D^2 h$$

$$= \left[-T + \sum_1^\infty \chi^m D_* T^{*m-1} D \right] Dh \qquad \text{(cf. (1.2.1))}.$$

Hence we deduce that

$$\Theta(\lambda) = \left[-T + \sum_1^\infty \lambda^m D_* T^{*m-1} D \right] \Bigg| \mathfrak{D}. \tag{3.2.7}$$

This operator $(\mathfrak{D} \to \mathfrak{D}_*)$ valued function is called the *characteristic function* of the contraction operator T; it will be denoted by $\Theta_T(\lambda)$ also. Note that

$$\Theta_T(0)d = -Td \quad \text{for } d \in \mathfrak{D}, \tag{3.2.8}$$

and hence $\|\Theta_T(0)d\| < \|d\|$ for $d \neq 0$.

Let us call a contractive analytic function $\{\mathfrak{A}, \mathfrak{A}_*, A(\lambda)\}$

(i) *pure* if $\|A(0)a\| < \|a\|$ for $a \in \mathfrak{A}$, $a \neq 0$;

(ii) *inner* if multiplication by the function $A(e^{it})$ is an isometry from $L^2(\mathfrak{A})$ into $L^2(\mathfrak{A}_*)$, or equivalently, if the values of this function are, a.e. on the unit circle, isometries from $\mathfrak{A}$ into $\mathfrak{A}_*$;

(iii) *outer* if multiplication by the function $A(e^{it})$ maps $H^2(\mathfrak{A})$ onto a dense linear manifold in $H^2(\mathfrak{A})$, i.e., if $\overline{AH^2(\mathfrak{A})} = H^2(\mathfrak{A}_*)$.

Theorem 1. *The characteristic function $\Theta_T(\lambda)$ of a cnu contraction operator T is a purely contractive analytic function. It is inner iff $T \in C_{\cdot 0}$, and outer iff $T \in C_{\cdot 1}$.*

The proof is based on properties (ι) and (κ) of Q.

3. For cnu T, we now use property (μ) to complete our construction of the functional model for T.

To this end, consider the selfadjoint operator valued function

$$\Delta_T(e^{it}) = [I - \Theta_T(e^{it})^* \Theta_T(e^{it})]^{1/2} \qquad (\mathfrak{D} \to \mathfrak{D})$$

on the unit circle. For $l \in M(\mathfrak{L})$ we have

$$\|(I-Q)l\|^2 = \|l\|^2 - \|Ql\|^2 = \|\Phi l\|^2 - \|\Phi_* Ql\|^2$$
$$= \|\Phi l\|^2 - \|\Theta_T \Phi l\|^2 = \|\Delta_T \Phi l\|^2.$$

Thus, taking the closure of the map

$$(I-Q)l \mapsto \Delta_T \Phi l \qquad (l \in M(\mathfrak{L}))$$

we get a unitary operator

$$\Phi_{\mathfrak{R}}: \mathfrak{R} \to \overline{\Delta_T L^2(\mathfrak{D})},$$

which also commutes with multiplication by χ (and hence by any scalar valued function in L^∞).

Then $\Psi = \Phi_* \oplus \Phi_{\mathfrak{R}}$ is a unitary operator

$$\Psi: \ \mathfrak{K} = M(\mathfrak{L}_*) \oplus \mathfrak{R} \to L^2(\mathfrak{D}_*) \oplus \overline{\Delta_T L^2(\mathfrak{D})} = \mathbf{K},$$

by which

$$\mathfrak{K}_+ = M_+(\mathfrak{L}_*) \oplus \mathfrak{R} \to H^2(\mathfrak{D}_*) \oplus \overline{\Delta_T L^2(\mathfrak{D})} = \mathbf{K}_+ \quad \text{and}$$
$$\mathfrak{H} = \mathfrak{K}_+ \ominus \{Ql \oplus (I-Q)l\colon \ l \in M_+(\mathfrak{L})\}$$
$$\to \mathbf{K}_+ \ominus \{\Theta_T u \oplus \Delta_T u\colon \ u \in H^2(\mathfrak{D})\} = \mathbf{H}.$$

Furthermore, Ψ transforms U to a unitary operator $\mathbf{U}$ on $\mathbf{K}$, $\mathbf{U} = \Psi U \Psi^{-1}$, such that $\mathbf{U}(u_* \oplus v) = \chi u_* \oplus \chi v$ ($u_* \oplus v \in \mathbf{K}$). The isometric dilation U_+ and T will therefore be transformed to the operators $\mathbf{U}_+ = \mathbf{U}|\mathbf{K}_+$ and $\mathbf{T}$; we have

$$\mathbf{T} = P_{\mathbf{H}} \mathbf{U}_+ | \mathbf{H} \quad \text{and} \quad \mathbf{T}^* = \mathbf{U}_+^* | \mathbf{H};$$

cf. property (γ). Thus we have, in particular,

Theorem 2. *Every cnu contraction operator T on a separable Hilbert space is unitarily equivalent to the operator $S(\Theta)$ on a Hilbert space $\mathfrak{H}(\Theta)$, associated with some purely contractive analytic function $\{\mathfrak{A}, \mathfrak{A}_*, \Theta(\lambda)\}$ in the following way:*

$$\mathfrak{H}(\Theta) = [H^2(\mathfrak{A}_*) \oplus \overline{\Delta L^2(\mathfrak{A})}] \ominus \{\Theta u \oplus \Delta u\colon \ u \in H^2(\mathfrak{A})\},$$
$$S(\Theta)(u_* \oplus v) = P_{\mathfrak{H}(\Theta)}(\chi u_* \oplus \chi v),$$
$$S(\Theta)^*(u_* \oplus v) = \overline{\chi}[u_* - u_*(0)] \oplus \overline{\chi} v \qquad (u_* \quad v \in \mathfrak{H}(\Theta)),$$ [3]

---

[3] $u_*(0) = (1/2\pi) \int_0^{2\pi} u_*(e^{it})dt.$

where $\Delta(e^{it}) = [I - \Theta(e^{it})^*\Theta(e^{it})]^{1/2}$. *For* $\Theta(\lambda)$ *we can choose, in particular, the characteristic function* $\{\mathfrak{D}, \mathfrak{D}_*, \Theta_T(\lambda)\}$ *defined by*

$$\Theta_T(\lambda) = \left[-T + \sum_1^\infty \lambda^m D_* T^{*m-1} D\right]\Big| \mathfrak{D}.$$

This function is inner iff $T \in C_{\cdot 0}$, *and outer iff* $T \in C_{\cdot 1}$.

This theorem has an important converse:

Theorem 3. *Given any purely contractive analytic function* $\{\mathfrak{A}, \mathfrak{A}_*, \Theta(\lambda)\}$, *the operator* $S(\Theta)$ *associated with it in the sense of the preceding theorem is a cnu contraction, whose characteristic function coincides with the given function* $\Theta(\lambda)$.

[Two operator valued functions, say $\{\mathfrak{A}, \mathfrak{A}_*, \Theta(\lambda)\}$ and $\{\mathfrak{A}', \mathfrak{A}'_*, \Theta'(\lambda)\}$ are said to coincide if there exist constant unitary operators $\alpha\colon \mathfrak{A} \to \mathfrak{A}'$, $\alpha_*\colon \mathfrak{A}_* \to \mathfrak{A}'_*$ such that $\Theta'(\lambda)\alpha = \alpha_*\Theta(\lambda)$.]

A possible proof of the theorem is to show first that $S(\Theta)$ is a cnu contraction and the operator $U(\Theta)$ defined on the space $K(\Theta) = L^2(\mathfrak{A}_*) \oplus \overline{\Delta L^2(\mathfrak{A})}$ by $U(\Theta)(u_* \oplus v) = \chi u_* \oplus \chi v$ is a minimal unitary dilation of $S(\Theta)$. Then determine the corresponding wandering subspaces $\mathfrak{L}$, $\mathfrak{L}_*$ which turn out to be in the following relation with $\mathfrak{A}$ and $\mathfrak{A}_*$:

$$\mathfrak{L} = \{\Theta a \oplus \Delta a\colon\ a \in \mathfrak{A}\}, \qquad \mathfrak{L}_* = \{a_* \oplus 0\colon\ a_* \in \mathfrak{A}_*\}.$$

Finally, determine the corresponding operator Q and the Fourier representations Φ, Φ_* in order to arrive at the characteristic function of $S(\Theta)$. We omit the details of these computations.

The two theorems together imply that *the operator* $S(\Theta)$ *associated with a purely contractive analytic function* $\Theta(\lambda)$ *is the general form, up to unitary equivalence, of a cnu contraction* T.

We know that $T \in C_{\cdot 0}$ if and only if Θ_T is an *inner* function, i.e., if $\Delta_T(e^{it}) = 0$ a.e. Hence, *the "functional model" of* $T \in C_{\cdot 0}$ *reduces to the particular form*

$$\begin{aligned} \mathfrak{H}(\Theta) &= H^2(\mathfrak{A}_*) \ominus \Theta H^2(\mathfrak{A}), \\ S(\Theta)u_* &= P_{\mathfrak{H}(\Theta)}(\chi u_*), \\ S(\Theta)^* u_* &= \bar{\chi}(u_* - u_*(0)), \end{aligned} \qquad (u_* \in \mathfrak{H}(\Theta)),$$

where $\{\mathfrak{A}, \mathfrak{A}_*, \Theta(\lambda)\}$ *is an inner function.*

4. It is not difficult to show that every contractive analytic function $\{\mathfrak{A}, \mathfrak{A}_*, \Theta(\lambda)\}$ is the direct sum of a purely contractive analytic function $\{\mathfrak{A}^\circ, \mathfrak{A}^\circ_*, \Theta^\circ(\lambda)\}$ and of a constant unitary valued function $\{\mathfrak{A}', \mathfrak{A}'_*, \Theta'(\lambda)\}$, that is,

$$\mathfrak{A} = \mathfrak{A}^\circ \oplus \mathfrak{A}', \qquad \mathfrak{A}_* = \mathfrak{A}^\circ_* \oplus \mathfrak{A}'_*,$$

$$\Theta^\circ(\lambda) = \Theta(\lambda)|\mathfrak{A}, \qquad \Theta'(\lambda) \equiv \Theta' = \Theta(\lambda)|\mathfrak{A}';$$

this decomposition is unique. One or another of the direct summands may be trivial ($\mathfrak{A}^\circ = \{0\}$ or $\mathfrak{A}' = \{0\}$).

The operator $S(\Theta)$ can be defined by the same formulas in the general case too, i.e., when $\Theta(\lambda)$ is not *purely* contractive. Then $S(\Theta)$ is still a cnu contraction, its characteristic function coincides with the purely contractive part $\Theta^\circ(\lambda)$ of $\Theta(\lambda)$.

In particular, the space $\mathfrak{H}(\Theta)$ is trivial ($= \{0\}$) iff $\Theta(\lambda)$ itself is a unitary valued constant.

4. Further Comments on the Characteristic Function $\Theta_T(\lambda)$

1. The characteristic function was defined by its power series expansion:

$$\Theta_T(\lambda) = \left[-T + \sum_{1}^{\infty} \lambda^m D_* T^{*m-1} D\right]\Bigg| \mathfrak{D}; \tag{4.1.1}$$

this definition, and the fact that $\Theta_T(\lambda)$ is contractive, followed from dilation theory.

A compact form can be given to $\Theta_T(\lambda)$ when multiplied from the right by D; as $D\mathfrak{H}$ is dense in $\mathfrak{D}$, $\Theta_T(\lambda)$ is defined once $\Theta_T(\lambda)D$ is defined. This compact form is the following:

$$\Theta_T(\lambda)D = D_*(I - \lambda T^*)^{-1}(\lambda I - T) \qquad (|\lambda| < 1). \tag{4.1.2}$$

From this formula we can get by straightforward calculation (and closure), the relation

$$\|f\|^2 - \|\Theta_T(\lambda)f\|^2 = (1 - |\lambda|^2)\|(I - \lambda T^*)^{-1}Df\|^2 \qquad (f \in \mathfrak{D}), \tag{4.1.3}$$

which shows that $\Theta_T(\lambda)$ is contractive. Also, we see that $\Theta_T(\lambda)$ is a fractional linear function of λ, with coefficients T and T^*.

From (4.1.1)—but from (4.1.2) also—it is easy to deduce the relation

$$\Theta_{T^*}(\lambda) = \Theta_T^{\sim}(\lambda) \tag{4.1.4}$$

where we use the notation $A^{\sim}(\lambda) = A(\bar{\lambda})^*$.

Again from (4.1.1) we obtain $\Theta_T(0) = -T|\mathfrak{D}$. Now, it is easy to prove that $Z = T|\mathfrak{D}^{\perp}$ is a unitary operator $\mathfrak{D}^{\perp} \to \mathfrak{D}_*^{\perp}$. Hence,

$$T = -\Theta_T(0) \oplus Z. \tag{4.1.5}$$

Now, elementary calculation shows that, for $|a| < 1$ and for any contraction T,

$$T_a = (T - aI)(I - \bar{a}T)^{-1} \tag{4.1.6}$$

is also a contraction, cnu if T is also. Moreover, using (4.1.2) one can show that

$$\Theta_{T_a}(\lambda) \textit{ coincides with } \Theta_T\left(\frac{\lambda + a}{1 + \bar{a}\lambda}\right);$$

hence, in particular,

$$\Theta_{T_a}(0) = V_a \Theta_T(a) W_a$$

for some unitary operators V_a, W_a. Applying (4.1.5) for T_a we get

$$T_a = -V_a \Theta_T(a) W_a \oplus Z_a \tag{4.1.7}$$

for some unitary Z_a.

As a consequence of (4.1.7) one sees that $\Theta_T(a)$ is *invertible* (i.e., injective) iff T_a is also and hence (by (4.1.6)) iff $T - aI$ is invertible, and that $\Theta_T(a)$ is *boundedly invertible* iff T_a is also and hence iff $T - aI$ is boundedly invertible. In the latter case,

$$\|\Theta_T(a)^{-1}\| = \|T_a^{-1}\| \begin{cases} \leq 1 + 2(1 - |\lambda|)\|(T - aI)^{-1}\|, \\ \geq (1 - |\lambda|)\|(T - aI)^{-1}\|. \end{cases} \tag{4.1.8}$$

In this way, the function $\Theta_T(\lambda)$ characterizes the points of the spectrum and of the point spectrum of T inside the unit circle.

It also characterizes the points of the spectrum of T *on* the unit circle. Notably, a point on the unit circle belongs to the resolvent set (i.e., to the complement of the spectrum) iff it lies in some open arc on which $\Theta_T(\lambda)$ is unitary operator valued and through which it can be extended analytically. That the points of the resolvent set of T lying on the circle have this property for $\Theta_T(\lambda)$ follows by straightforward computation, while for the converse statement one uses a kind of Schwarz reflection principle (for details see [H, Chapter VI.4]).

2. One of the interesting uses of the characteristic function $\Theta_T(\lambda)$ is that it yields a criterion for the contraction T to be *similar* to some *unitary* operator V. Similarity means that there exists a boundedly invertible operator S (i.e., an "affinity") such that

$$T = S^{-1} V S. \tag{4.2.1}$$

The first part of the reasoning is of geometrical character: it involves the minimal isometric dilation U_+ of T.

From §2.1 we know that if $\mathfrak{R}$ denotes the space of the unitary part of U_+ then

$$P_{\mathfrak{R}} h = \lim_{n\to\infty} U_+^n T^{*n} h \quad \text{for } h \in \mathfrak{H}; \tag{4.2.2}$$

whence,

$$U_+^n P_{\mathfrak{R}} T^{*n} h = P_{\mathfrak{R}} h \quad \text{for } h \in \mathfrak{H} \text{ and } n = 0, 1, \cdots. \tag{4.2.3}$$

Suppose (4.2.1) holds. Then we have $T^{*n} = S^* V^{*n} S^{*-1}$, and therefore (4.2.2) implies

(4.2.4) $\|P_{\mathfrak{R}}h\| \geq c\|h\|$ for all $h \in \mathfrak{H}$, with some $c > 0$,

namely with $1/c = \|S^*\|\|S^{*-1}\| = \|S\|\|S^{-1}\|$. Moreover, we then have $T^{*n}\mathfrak{H} = \mathfrak{H}$ for $n \geq 0$. Also using (4.2.3) and recalling that U_+ commutes with $P_{\mathfrak{R}}$ we obtain

$$\mathfrak{R} = P_{\mathfrak{R}}\mathfrak{K}_+ = P_{\mathfrak{R}} \bigvee_{n=0}^{\infty} U_+^n \mathfrak{H} = \bigvee_{n=0}^{\infty} U_+^n P_{\mathfrak{R}}\mathfrak{H}$$

$$= \bigvee_{n=0}^{\infty} U_+^n P_{\mathfrak{R}} T^{*n}\mathfrak{H} = \overline{P_{\mathfrak{R}}\mathfrak{H}}.$$

Observe that (4.2.4) implies that $P_{\mathfrak{R}}\mathfrak{H}$ is closed. Thus, if T is similar to a unitary operator then

$$(*) \qquad \|P_{\mathfrak{R}}h\| \geq c\|h\| \text{ for all } h \in \mathfrak{H}, \text{ with some constant } c > 0; \quad \text{and} \quad P_{\mathfrak{R}}\mathfrak{H} = \mathfrak{R}.$$

Conversely, if (*) holds then $P_{\mathfrak{R}}|\mathfrak{H}$ is an affinity from $\mathfrak{H}$ onto $\mathfrak{R}$, and therefore its adjoint X is an affinity from $\mathfrak{R}$ onto $\mathfrak{H}$; moreover (4.2.3) implies that $R(P_{\mathfrak{R}}|\mathfrak{H})T^* = P_{\mathfrak{R}}|\mathfrak{H}$, and hence $TXR^* = X$, where R denotes the unitary part of U_+. Thus if (*) holds then T is similar to the unitary operator R.

We conclude that condition (*) is necessary and sufficient for T to be similar to a unitary operator.

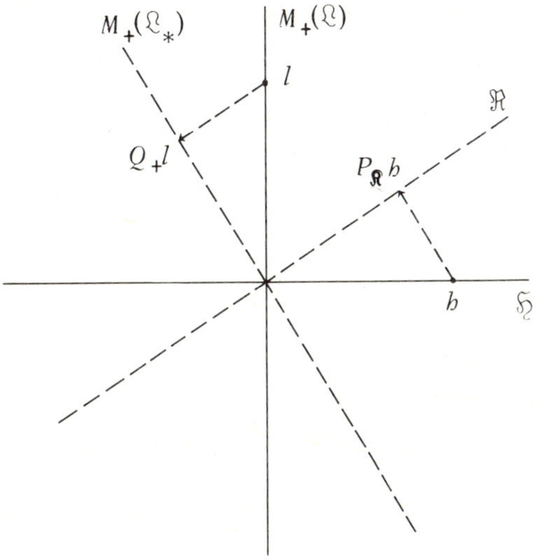

Now, using the two decompositions of $\mathfrak{K}_+$,

$$\mathfrak{K}_+ = \mathfrak{H} \oplus M_+(\mathfrak{L}), \; \mathfrak{K}_+ = M_+(\mathfrak{L}_*) \oplus \mathfrak{R},$$

(cf. (1.4.4) and (1.4.5)) and denoting by Q_+ the orthogonal projection from $M_+(\mathfrak{L})$ into $M_+(\mathfrak{L}_*)$, it is not difficult to prove that condition (*) is equivalent to the condition

$$(**) \qquad \|Q_+l\| \geq c\|l\| \text{ for all } l \in M_+(\mathfrak{L}), \quad Q_+M_+(\mathfrak{L}) = M_+(\mathfrak{L}_*),$$

with the same constant $c > 0$.

As Q_+ is the restriction to $M_+(\mathfrak{L})$ of the operator Q considered in §2, we can translate condition (**) by using the Fourier transformations Φ and Φ_* introduced in §3.2 (or rather their restrictions to $M_+(\mathfrak{L})$ and $M_+(\mathfrak{L}_*)$) to arrive at the result that condition (**) is equivalent to the condition that the operator Θ_T: $H^2(\mathfrak{D}) \to H^2(\mathfrak{D}_*)$

is onto and has a bounded inverse; and this in turn is equivalent to the condition that the values of the characteristic function $\Theta_T(\lambda)$ $(|\lambda| < 1)$, as operators $\mathfrak{D} \to \mathfrak{D}_*$, are onto and invertible with the bound $1/c$ for $\Theta_T(\lambda)^{-1}$.

This concludes the proof of the following:

Theorem. *The contraction T is similar to a unitary operator iff*

$$\|\Theta_T(\lambda)^{-1}\| \leq C \qquad (|\lambda| < 1),$$

with some constant C independent of λ.

This criterion (Sz.-Nagy-Foiaş, [H, Chapter IX.1]) can be given because of the estimates (4.1.8), the following equivalent form (Gohberg-Krein):

The contraction T is similar to a unitary operator iff there is a constant C' such that

$$\|(T - \lambda I)^{-1}\| \leq C'(1 - |\lambda|) \quad \text{for } |\lambda| < 1.$$

It is remarkable that—although in this form neither the characteristic function nor the unitary dilation is involved—there does not exist as yet a direct proof of the criterion.

5. Invariant Subspaces and Factorizations of the Characteristic Function

1. As we already know, the general form (up to unitary equivalence) of a contraction T of class $C_{\cdot 0}$ (on a separable Hilbert space) is the operator $S(\Theta)$ defined on the space $\mathfrak{H}(\Theta)$ corresponding to a purely contractive *inner* function $\{\mathfrak{A}, \mathfrak{A}_*, \Theta(\lambda)\}$; that is,

$$\mathfrak{H}(\Theta) = H^2(\mathfrak{A}_*) \ominus \Theta H^2(\mathfrak{A}),$$

$$S(\Theta)u_* = P_{\mathfrak{H}(\Theta)}(\chi u_*), \quad S(\Theta)^* u_* = \bar{\chi}(u_* - u_*(0)), \qquad \text{for } u_* \in \mathfrak{H}(\Theta).$$

Let $\Theta(\lambda) = \Theta_2(\lambda)\Theta_1(\lambda)$ be a factorization of $\Theta(\lambda)$ into inner factors, say $\{\mathfrak{A}, \mathfrak{B}, \Theta_1(\lambda)\}$ and $\{\mathfrak{B}, \mathfrak{A}_*, \Theta_2(\lambda)\}$:

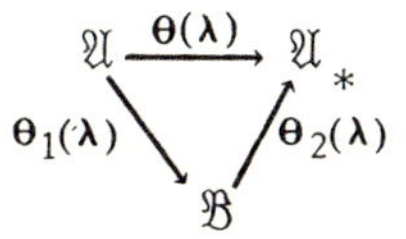

This factorization generates the following orthogonal decomposition of $\mathfrak{H}(\Theta)$:

$$\begin{aligned}\mathfrak{H}(\Theta) &= H^2(\mathfrak{A}_*) \ominus \Theta_2\Theta_1 H^2(\mathfrak{A}) \\ &= \Theta_2[H^2(\mathfrak{B}) \ominus \Theta_1 H^2(\mathfrak{A})] \oplus [H^2(\mathfrak{A}_*) \ominus \Theta_2 H^2(\mathfrak{B})] \\ &= \Theta_2\mathfrak{H}(\Theta_1) \oplus \mathfrak{H}(\Theta_2) \equiv \mathfrak{H}_1 \oplus \mathfrak{H}_2.\end{aligned}$$

For $u_1 \in \mathfrak{H}(\Theta_1)$ we have

$$S(\Theta)\Theta_2 u_1 = P_{\mathfrak{H}(\Theta)}(\chi\Theta_2 u_1) = P_{\mathfrak{H}_1}(\Theta_2\chi u_1),$$

because $\Theta_2\chi u_1 \in \Theta_2 H^2(\mathfrak{B}) \perp \mathfrak{H}_2$. Now as $\Theta_2(\lambda)$ is an inner function $\omega_1 = \Theta_2|\mathfrak{H}(\Theta_1)$ maps $\mathfrak{H}(\Theta_1)$ unitarily onto $\mathfrak{H}_1$ and so we have

$$P_{\mathfrak{H}_1}(\Theta_2\chi u_1) = P_{\omega_1\mathfrak{H}(\Theta_1)}(\omega_1\chi u_1) = \omega_1 P_{\mathfrak{H}(\Theta_1)}(\chi u_1) = \omega_1 S(\Theta_1)u_1;$$

we conclude that

$$S(\Theta)\omega_1 = \omega_1 S(\Theta_1).$$

Thus $\mathfrak{H}_1$ is an invariant subspace for $S(\Theta)$, and the restriction $S(\Theta)|\mathfrak{H}_1$ is unitarily equivalent to $S(\Theta_1)$.

Then $\mathfrak{H}_2$ is invariant for $S(\Theta)^*$. For $u \in \mathfrak{H}_2$ we have

$$S(\Theta)^* u = \overline{\chi}(u - u_0) = S(\Theta_2)^* u;$$

therefore $S(\Theta)^*|\mathfrak{H}_2 = S(\Theta_2)^*$.

Thus the operator $S(\Theta)$ has, with respect to the decomposition $\mathfrak{H}(\Theta) = \mathfrak{H}_1 \oplus \mathfrak{H}_2$, a matrix of the form

$$\begin{bmatrix} S_1 & X \\ 0 & S_2 \end{bmatrix}$$

where S_1 is unitarily equivalent to $S(\Theta_1)$, while S_2 equals $S(\Theta_2)$. Hence the characteristic functions of S_1 and S_2 coincide with the pure parts of $\Theta_1(\lambda)$ and $\Theta_2(\lambda)$. It follows in particular that *if the factorization is nontrivial* (i.e., none of the factors $\Theta_1(\lambda)$, $\Theta_2(\lambda)$ is a unitary constant) *then* $\mathfrak{H}_1$ *is a proper invariant subspace for* $S(\Theta)$.

Moreover, one thus obtains *all* subspaces $\mathfrak{H}_1$ invariant for $S(\Theta)$, or equivalently, *all* subspaces $\mathfrak{H}_2$ invariant for $S(\Theta)^*$. Indeed, if $\mathfrak{H}_2$ is any invariant subspace for $S(\Theta)^*$ then it is also invariant for the backward shift S^*: $u \mapsto \overline{\chi}(u - u_0)$ on $H^2(\mathfrak{A}_*)$, and therefore $H^2(\mathfrak{A}_*) \ominus \mathfrak{H}_2$ is invariant for the unilateral (forward) shift S: $u \mapsto \chi u$ on $H^2(\mathfrak{A}_*)$. Now by a theorem of Beurling-Lax-Halmos such a subspace must be of the form $\Theta_2 H^2(\mathfrak{B})$, where $\{\mathfrak{B}, \mathfrak{A}_*, \Theta_2(\lambda)\}$ is some inner function. Hence,

$$\mathfrak{H}_2 = H^2(\mathfrak{A}_*) \ominus \Theta_2 H^2(\mathfrak{B}) = \mathfrak{H}(\Theta_2).$$

As $\Theta_2 H^2(\mathfrak{B}) \supset \Theta_2 \Theta_1 H^2(\mathfrak{A}) = \Theta H^2(\mathfrak{A})$ and as both $\Theta(\lambda)$ and $\Theta_1(\lambda)$ are inner, we can easily conclude, by using a lemma on operators commuting with multiplication by χ, that $\Theta(\lambda) = \Theta_2(\lambda)\Theta_1(\lambda)$ with some inner function $\{\mathfrak{A}, \mathfrak{B}, \Theta_1(\lambda)\}$.

Thus for $T \in C_{\cdot 0}$ there is a complete correspondence between the invariant subspaces for T and the inner function factorizations of $\Theta_T(\lambda)$. Unfortunately, the problem of finding the inner function factorizations of an inner function $\Theta(\lambda)$ has been (to our knowledge) solved only for $\Theta(\lambda)$ possessing a *scalar multiple*.

2. Let us now consider cnu contractions T of general type, or equivalently, operators $S(\Theta)$ corresponding to an arbitrary contractive analytic function $\{\mathfrak{A}, \mathfrak{A}_*, \Theta(\lambda)\}$.

Observe that if $\Theta(\lambda) = \Theta_2(\lambda)\Theta_1(\lambda)$ is any factorization of $\Theta(\lambda)$ with factors of the same type then—using the function $\Delta(e^{it}) = [I - \Theta(e^{it})^*\Theta(e^{it})]^{1/2}$ and its analogs—we get

$$\Delta(\cdot)^2 = \Theta_2(\cdot)^*\Delta_2(\cdot)^2\Theta_2(\cdot) + \Delta_1(\cdot)^2. \tag{5.2.1}$$

This implies that the transformation

$$\Delta(e^{it})a \mapsto \Delta_2(e^{it})\Theta_1(e^{it})a \oplus \Delta_1(e^{it})a \qquad (a \in \mathfrak{A})$$

is isometric for almost every (fixed) point e^{it}, and hence extends by closure to an isometry

$$Z(e^{it}):\ \overline{\Delta(e^{it})\mathfrak{A}} \to \overline{\Delta_2(e^{it})\mathfrak{B}} \oplus \overline{\Delta_1(e^{it})\mathfrak{A}}.$$

From (5.2.1) it also follows that the transformation

$$\Delta u \mapsto \Delta_2\Theta_1 u \oplus \Delta_1 u \qquad (u \in L^2(\mathfrak{A}))$$

is also isometric, and hence extends by closure to an isometry

$$Z:\ \overline{\Delta L^2(\mathfrak{A})} \to \overline{\Delta_2 L^2(\mathfrak{B})} \oplus \overline{\Delta_1 L^2(\mathfrak{A})}.$$

It is easy to show that the following two conditions are equivalent:

Local condition: $Z(e^{it})$ is unitary (i.e., onto) for almost every fixed point e^{it}.

Global condition: Z is unitary (i.e., onto).

Whenever these (equivalent) conditions are satisfied we call the factorization $\Theta(\lambda) = \Theta_2(\lambda)\Theta_1(\lambda)$ *regular.* If this is the case, we can identify the elements of the spaces

$$\overline{\Delta L^2(\mathfrak{A})} \quad \text{and} \quad \overline{\Delta_2 L^2(\mathfrak{B})} \oplus \overline{\Delta_1 L^2(\mathfrak{A})}$$

which correspond by the unitary operator Z, thus obtaining for $\mathfrak{H}(\Theta)$ the following three-component form:

$$\mathfrak{H}(\Theta) = \left[H^2(\mathfrak{A}_*) \oplus \overline{\Delta_2 L^2(\mathfrak{B})} \oplus \overline{\Delta_1 L^2(\mathfrak{A})}\right]$$
$$\ominus \{\Theta_2\Theta_1 u \oplus \Delta_2\Theta_1 u \oplus \Delta_1 u:\ u \in H^2(\mathfrak{A})\},$$

and hence,

$$\mathfrak{H}(\Theta) = \mathfrak{H}_1 \oplus \mathfrak{H}_2,$$

where

$$\mathfrak{H}_2 = \left[H^2(\mathfrak{A}_*) \oplus \overline{\Delta_2 L^2(\mathfrak{B})} \oplus \overline{\Delta_1 L^2(\mathfrak{A})}\right]$$
$$\ominus \left\{\Theta_2 w \oplus \Delta_2 w \oplus v:\ w \in H^2(\mathfrak{B}),\ v \in \overline{\Delta_1 L^2(\mathfrak{A})}\right\}$$
$$= \mathfrak{H}(\Theta_2) \oplus \{0\}$$

and

$$\begin{aligned} \mathfrak{H}_1 &= \left\{\Theta_2 w \oplus \Delta_2 w \oplus v:\ w \in H^2(\mathfrak{B}),\ v \in \overline{\Delta_1 L^2(\mathfrak{A})}\right\} \\ &\ominus \{\Theta_2\Theta_1 u \oplus \Delta_2\Theta_1 u \oplus \Delta_1 u:\ u \in H^2(\mathfrak{A})\} \\ &= \omega_2 \cdot \mathfrak{H}(\Theta_1); \end{aligned} \tag{5.2.2}$$

ω_2 denotes the unitary operator $\mathfrak{H}(\Theta_1) \to \mathfrak{H}_1$ which results by restriction to $\mathfrak{H}(\Theta_1)$ of the isometry

$$u \oplus v \mapsto \Theta_2 u \oplus \Delta_2 u \oplus v \qquad (u \in H^2(\mathfrak{B}),\ v \in L^2(\mathfrak{A})).$$

Arguing as in the case of inner factorizations we get that $\mathfrak{H}_1$ is invariant for $S(\Theta)$ and in the corresponding matrix representation

$$S(\Theta) = \begin{bmatrix} S_1 & X \\ 0 & S_2 \end{bmatrix} \tag{5.2.3}$$

the operators S_1, S_2 are unitarily equivalent respectively to $S(\Theta_1)$ and $S(\Theta_2)$. In particular, $\mathfrak{H}_1$ is a *proper* invariant subspace for $S(\Theta)$ iff the factorization is *nontrivial.*

These arguments were quite simple. It is more difficult to prove that—in this general case too—we obtain in such a way *all* invariant subspaces for $S(\Theta)$. The proof uses a more detailed study of the structure of the minimal isometric dilation and, in fact, yields a substantial generalization of the Beurling-Lax-Halmos theorem used in §1.1. We do not go into details.

Thus we have the following:

Theorem. *As* $\Theta(\lambda) = \Theta_2(\lambda)\Theta_1(\lambda)$ *runs over all regular factorizations of the given contractive analytic* $\Theta(\lambda)$, *the subspace* $\mathfrak{H}_1$ *of* $\mathfrak{H}(\Theta)$ *given by formula* (5.2.2) *runs over all invariant subspaces for* $S(\Theta)$. *Moreover, in the corresponding triangularization*

$$S(\Theta) = \begin{bmatrix} S_1 & * \\ 0 & S_2 \end{bmatrix},$$

the operators S_1 *and* S_2 *are unitarily equivalent to* $S(\Theta_1)$ *and* $S(\Theta_2)$, *respectively.*

3. This theorem does not exclude the possibility that an invariant subspace with the above properties could be attached in some way (if not by means of (5.2.2)) to certain nonregular factorizations also. Such "strange" factorizations do exist; cf. Foiaş [3]. However, not all nonregular factorizations are strange. For example, if $\Theta(\lambda) = \Theta_2(\lambda)\Theta_1(\lambda)$ is any factorization of an inner function $\Theta(\lambda)$ and if

$$S(\Theta) = \begin{bmatrix} S_1 & * \\ 0 & S_2 \end{bmatrix}$$

is a triangulation with S_1, S_2 unitarily equivalent to $S(\Theta_1)$, $S(\Theta_2)$, respectively, then $S(\Theta) \in C_{\cdot 0}$ implies $S(\Theta_1)$, $S(\Theta_2) \in C_{\cdot 0}$ and therefore $\Theta_1(\lambda)$, $\Theta_2(\lambda)$ are also inner. As inner factorizations are all regular, there exists no strange factorizations for an inner function. But on the other hand we know that if the underlying Hilbert space $\mathfrak{A}$ is infinite dimensional, the inner function $\Theta(\lambda)$ does have nonregular factorizations, even such that $\Theta_1(\lambda)$ is inner while $\Theta_2(\lambda)$ is outer (and neither a unitary constant).

At the present moment we do not possess a means to determine all contractive analytic functions $\Theta(\lambda)$ which possess strange factorizations—and the strange factorizations themselves.

6. Commutative Systems

1. A natural generalization of the comcept of dilation of a single operator is the following one:

Let $\mathfrak{A} = \{A_\omega\}$ be a commutative system of operators on a Hilbert space $\mathfrak{H}$. We say that a corresponding system $\mathfrak{B} = \{B_\omega\}$ of operators on a Hilbert space $\mathfrak{K}$ is a *dilation* of $\mathfrak{A}$ if $\mathfrak{K}$ contains $\mathfrak{H}$ as a subspace, $\mathfrak{B}$ is commutative, and

$$A_{\omega_1}^{n_1} \cdots A_{\omega_r}^{n_r} = P_{\mathfrak{H}} B_{\omega_1}^{n_1} \cdots B_{\omega_r}^{n_r} | \mathfrak{H} \qquad (n_i \geq 0;\ i = 1, \cdots, r) \tag{6.1.1}$$

for every finite subset $\{\omega_1, \cdots, \omega_r\}$ of the set of subscripts.

Theorem (T. Ando). *Every commuting pair* $\{T_1, T_2\}$ *of contractions has a unitary dilation* $\{U_1, U_2\}$ (*i.e., composed of unitary operators* U_1, U_2).

Since it is easy to show that any pair of commuting isometric operators can be extended to a pair of commuting unitary operators (the same fact holds, by the way, for any number of commuting isometries as well) it suffices to prove that a dilation $\{V_1, V_2\}$ exists, with V_1, V_2 *isometries.*

Without aiming at any minimality property in the construction we consider the Hilbert space $\mathfrak{K}$ of vectors

$$k = (h_0, h_1, h_2, \cdots), \quad \text{with } h_i \in \mathfrak{H},\ \|k\|^2 = \sum_0^\infty \|h_i\|^2 \leq \infty,$$

in which $\mathfrak{H}$ will be embedded through the identification

$$h \sim (h, 0, 0, \cdots) \qquad (h \in \mathfrak{H}).$$

Next we define the isometries W_i $(i = 1, 2)$ on $\mathfrak{K}$ by

$$W_i(h_0, h_1, h_2, \cdots) = (T_i h_0, D_i h_0, 0, h_1, h_2, \cdots),$$

where $D_i = (I - T_i^* T_i)^{1/2}$. Then we obviously have

$$T_1^{n_1} T_2^{n_2} = P_{\mathfrak{H}} W_1^{n_1} W_2^{n_2} | \mathfrak{H} \qquad (n_i \geq 0)$$

(even if T_1, T_2 do not commute); but W_1, W_2 do not commute in general (even if T_1, T_2 do). Thus the problem is to modify the pair W_1, W_2 without changing the 0th components, so that the new pair are commuting isometries.

To this effect we consider an auxiliary unitary operator G on the space $\mathfrak{G} = \mathfrak{H} \oplus \mathfrak{H} \oplus \mathfrak{H} \oplus \mathfrak{H}$, to be determined later, and the unitary operator $\mathbf{G}$ on $\mathfrak{K}$ defined by

$$\mathbf{G}k = \mathbf{G}\langle h_0, h_1, h_2, \cdots\rangle = \langle h_0, G\langle h_1, \cdots, h_4\rangle, G\langle h_5, \cdots, h_8\rangle, \cdots\rangle.$$

Then we consider the isometries $V_1 = \mathbf{G}W_1$ and $V_2 = W_2\mathbf{G}^{-1}$; it is obvious that they are admissible modifications of W_1 and W_2. An easy calculation shows

$$V_1V_2\langle h_0, h_1, \cdots\rangle = \langle T_1T_2h_0, G\langle D_1T_2h_0, 0, D_2h_2, 0\rangle, \langle h_1, \cdots, h_4\rangle, \cdots\rangle,$$

$$V_2V_1\langle h_0, h_1, \cdots\rangle = \langle T_2T_1h_0, D_2T_1h_0, 0, D_1h_0, 0, h_1, h_2, \cdots\rangle.$$

Thus V_1 and V_2 commute if and only if

$$G\langle D_1T_2h, 0, D_2h, 0\rangle = \langle D_2T_1h_0, 0, D_1h, 0\rangle$$

for every $h \in \mathfrak{H}$. As we have

$$\|\langle D_1T_2h, 0, D_2h, 0\rangle\|^2 = \|h\|^2 - \|T_1T_2h\|^2 = \|\langle D_2T_1h, 0, D_1h, 0\rangle\|^2$$

(it is at this point that we use the commutativity of T_1 and T_2), the above condition fixes G as an isometry from a linear manifold of $\mathfrak{G}$, namely from $\mathfrak{G}_1 = \{\langle D_1T_2h, 0, D_2h, 0\rangle\}$ onto $\mathfrak{G}_2 = \{\langle D_2T_1h, 0, D_1h, 0\rangle\}$. In order that G can be extended to a unitary operator on $\mathfrak{G}$, it is thus necessary and sufficient that the orthogonal complements $\mathfrak{G}_1^\perp$ and $\mathfrak{G}_2^\perp$ (in $\mathfrak{G}$) be equidimensional. If $\dim \mathfrak{G}$ ($= 4\dim \mathfrak{H}$) $< \infty$, this equidimensionality is obvious. If $\dim \mathfrak{H} = \infty$, then we observe that both $\mathfrak{G}_1^\perp$ and $\mathfrak{G}_2^\perp$ contain subspaces of the same dimension as $\mathfrak{H}$; hence

$$\dim \mathfrak{H} \leq \dim \mathfrak{G}_i^\perp \leq \dim \mathfrak{G} = 4 \dim \mathfrak{H} = \dim \mathfrak{H} \qquad (i = 1, 2),$$

and therefore we have here equality, and the proof is complete.

2. Although the above proof is essentially dependant on the number 2 of the contractions involved, for some years it was hoped that the theorem would turn out to hold for systems of more than two operators also.

These hopes were killed by a counterexample found by S. Parrott, for three contractions. Indeed, *there even exist such commuting contractions* T_1, T_2, T_3 *for which we cannot find commuting unitaries* U_1, U_2, U_3 *such that* $T_i = P_{\mathfrak{H}}U_i|\mathfrak{H}$ $(i = 1, 2, 3)$.

For the construction of such a counterexample let us start with any three unitary operators A_1, A_2, A_3 on a Hilbert space $\mathfrak{A}$, such that

$$A_1A_2^{-1}A_3 \neq A_3A_2^{-1}A_1. \qquad (6.2.1)$$

(E.g., choose $A_2 = I$, and, for A_1, A_3, two noncommuting unitary operators, which is always possible if dim $\mathfrak{A} \geq 2$.) Let $\mathfrak{H} = \mathfrak{A} \oplus \mathfrak{A}$ and define T_i on $\mathfrak{H}$ by

$$T_i\langle a_1, a_2\rangle = \langle 0, A_i a_1\rangle \qquad (i = 1, 2, 3).$$

Clearly $\|T_i\| = 1$ and $T_i T_j = 0$ for $i, j = 1, 2, 3$. Suppose there exist commuting unitaries U_i $(i = 1, 2, 3)$, with $T_i = P_{\mathfrak{H}} U_i|\mathfrak{H}$. Then

$$P_{\mathfrak{H}} U_i\langle a, 0\rangle = T_i\langle a, 0\rangle = \langle 0, A_i a\rangle, \qquad \|U_i\langle a, 0\rangle\| = \|\langle a, 0\rangle\| = \|\langle 0, A_i a\rangle\|;$$

therefore U_i $a, 0$ $=$ $0, A_i a$, and hence

$$U_j^{-1}U_i\langle a, 0\rangle = \langle A_j^{-1}A_i a, 0\rangle, \qquad U_k U_j^{-1} U_i\langle a, 0\rangle = \langle 0, A_k A_j^{-1} A_i a\rangle.$$

Since the U's commute we conclude that $A_k A_j^{-1} A_i = A_i A_j^{-1} A_k$, and this contradicts (6.2.1).

3. Let $\mathfrak{A} = \{A_\omega\}$ be a commutative system of operators on a Hilbert space $\mathfrak{H}$ and let $\mathfrak{B} = \{B_\omega\}$ be a dilation of $\mathfrak{A}$. This dilation is called *regular* if the relation

$$(6.3.1) \qquad \prod_i A_{\omega_i'}^{*m_i} \cdot \prod_j A_{\omega_j}^{n_j} = P_{\mathfrak{H}} \prod_i B_{\omega_i'}^{*m_i} \cdot \prod_j B_{\omega_j}^{n_j}\Big|\mathfrak{H}$$

holds for any two, disjoint finite subsets $\{\omega_j\}$ and $\{\omega_i'\}$ of the set of subscripts ω, and any integers $m_i, n_j \geq 0$.

Although we do not possess any useful characterizations of commutative systems of (more than two) contractions admitting unitary dilations, we do have such a characterization of commutative systems of (two or more) contractions which admit a unitary *regular* dilation. Notably, we have

Theorem (Brehmer). *In order that the commutative system $\mathfrak{T} = \{T_\omega\}$ of contractions on $\mathfrak{H}$ admit a unitary, regular dilation it is necessary and sufficient that the inequality*

$$(6.3.2) \qquad \|h\|^2 - \sum_{1\leq i\leq r} \|T_i h\|^2 + \sum_{1\leq i<j\leq r} \|T_i T_j h\|^2 - \sum_{1\leq i<j<k\leq r} \|T_i T_j T_k h\|^2 + \cdots \pm \|T_1 T_2 \cdots T_r h\|^2 \geq 0$$

hold for every finite subset $\{T_1, T_2, \cdots, T_r\}$ of $\mathfrak{T}$ and for every $h \in \mathfrak{H}$.

Special cases.

(1) $\mathfrak{T}$ consists of isometries. In this case the left-hand side of (6.3.2) is equal to $(1 - 1)^r\|h\|^2$, i.e., to 0.

(2) $\mathfrak{T}$ consists of *doubly* commuting contractions, i.e., T_ω commutes with $T_{\omega'}$ and with $T_{\omega'}^*$, for $\omega \neq \omega'$. For, in this case, the left-hand side of (6.3.2) is equal to

$\Pi_{1 \le i \le r} (I - T_i^* T_i)$, and this product is a positive operator since its factors are positive and commute.

(3) $\Sigma_\omega \|T_\omega\|^2 \le 1$. Then set, for $p = 0, 1, \cdots, r$,

$$a_p(h) = \sum_{1 \le i_1 < \cdots < i_p \le r} \|T_{i_1} \cdots T_{i_p} h\|^2,$$

and observe that, for $p = 1, \cdots, r$,

$$a_p(h) \le \sum_{1 \le i_1 < \cdots < i_p \le r} \|T_{i_p}\|^2 \|T_{i_1} \cdots T_{i_{p-1}} h\|^2$$

$$= \sum_{1 \le i_1 < \cdots < i_{p-1} < r} \left[\|T_{i_1} \cdots T_{i_{p-1}} h\|^2 \sum_{i_{p-1} < i \le r} \|T_i\|^2 \right] \le a_{p-1}(h),$$

and hence $a_0(h) - a_1(h) + a_2(h) - \cdots \pm a_r(h) \ge 0$.

4. If a finite commutative system $\{T_1, \cdots, T_r\}$ of contractions on $\mathfrak{H}$ has a unitary dilation $\{U_1, \cdots, U_r\}$ then we have

$$p(T_1, \cdots, T_r) = P_{\mathfrak{H}} p(U_1, \cdots, U_r) | \mathfrak{H}$$

for every polynomial $p(\lambda_1, \cdots, \lambda_r)$, and hence

$$\|p(T_1, \cdots, T_r)\| \le \|p(U_1, \cdots, U_r)\|.$$

From the spectral theory for commuting unitary operators it follows that $\|p(U_1, \cdots, U_r)\| \le \|p\|_\infty$, where $\|p\|_\infty$ denotes the supremum of the function $|p(\lambda_1, \cdots, \lambda_r)|$ on the domain $|\lambda_i| \le 1$ $(i = 1, \cdots, r)$. Thus we have then

$$\|p(T_1, \cdots, T_r)\| \le \|p\|_\infty. \tag{6.4.1}$$

Now as any single contraction and any commuting pair of contractions have unitary dilations, inequality (6.4.1) holds if $r = 1$ or $r = 2$ (for $r = 1$ it was first proved, by a quite different method, by J. von Neumann, 1951).

Quite recently, N. Varopoulos [15] proved that (6.4.1) is not true for *all* commutative systems of contractions. There are counterexamples even for $r = 3$.

7. Lifting of Intertwining Operators

1. Suppose T is a contraction on $\mathfrak{H}$ and U_+ is its minimal isometric dilation on the space

$$\mathfrak{K}_+ = \mathfrak{H} \oplus \mathfrak{L} \oplus U_+\mathfrak{L} \oplus U_+^2\mathfrak{L} \oplus \cdots;$$

cf. (1.4.4). Denote by P_+ the projection of $\mathfrak{K}_+$ onto $\mathfrak{H}$. Finally, let W be an isometry on some space $\mathfrak{G}$.

Theorem 1. *For every operator*

(7.1.1) $$X\colon \mathfrak{G} \to \mathfrak{H} \text{ such that } TX = XW$$

there exists an operator

(7.1.2) $$Y\colon \mathfrak{G} \to \mathfrak{K}_+ \text{ such that } U_+Y = YW, \; X = P_+Y;$$

moreover we can require that

(7.1.3) $$\|Y\| = \|X\|.$$

Proof. It is enough to consider the case $\|X\| = 1$ and then construct Y satisfying (7.1.2) and $\|Y\| \le 1$. (Indeed, $\|X\| \le \|Y\|$ is a trivial consequence of $X = P_+Y$.)

We have then to choose Y in the form

(7.1.4) $$Y = X + B_0 + U_+B_1 + U_+^2B_2 + \cdots$$

with operators $B_n\colon \mathfrak{G} \to \mathfrak{L}$. To get $\|Y\| \le 1$ we must have

(a) $$\|Xg\|^2 + \sum_0^\infty \|B_ng\|^2 \le \|g\|^2 \quad \text{for all } g \in \mathfrak{G}.$$

Because (7.1.4) implies

$$U_+Y - YW = \sum_0^\infty U_+^n(B_{n-1} - B_nW) \quad \text{with } B_{-1} = U_+X - XW,$$

we must have, on account of (7.1.1) and (7.1.2),

(b) $$B_{n-1} = B_nW \quad (n \ge 0) \qquad \text{and} \qquad B_{-1} = (U_+ - T)X.$$

From (7.1.1), the isometry of W, and the assumption $\|X\| = 1$ we infer

$$\begin{aligned}\|B_{-1}g\|^2 &= \|(U_+ - T)Xg\|^2 = \|Xg\|^2 - \|TXg\|^2 \le \|g\|^2 - \|XWg\|^2 \\ &= \|Wg\|^2 - \|XWg\|^2,\end{aligned} \tag{7.1.5}$$

and hence

$$\|B_{-1}g\|^2 \le \|D_0Wg\|^2, \quad \text{where } D_0 = (I_{\mathfrak{G}} - X^*X)^{1/2}.$$

Thus the map $C_0\colon D_0Wg \mapsto B_{-1}g$ $(g \in \mathfrak{G})$ is contractive and linear, and therefore can be extended to a contraction of the whole space $\mathfrak{G}$ into $\mathfrak{K}$; let us denote this extension by the same letter C_0. Define $B_0 = C_0D_0$ then, obviously, $B_0W = B_{-1}$, i.e., (b) holds for $n = 0$ and, moreover,

$$\|Xg\|^2 + \|B_0g\|^2 = \|Xg\|^2 + \|C_0D_0g\|^2 \le \|Xg\|^2 + \|D_0g\|^2 = \|g\|^2.$$

We proceed by recurrence. Suppose the operators B_n with $n < N$ (where $N \ge 1$) are already determined and satisfy the conditions

$$(\mathrm{a})_N \qquad s_N(g) := \|Xg\|^2 + \sum_0^{N-1} \|B_ng\|^2 \le \|g\|^2 \qquad (g \in \mathfrak{G}),$$

$$(\mathrm{b})_N \qquad B_nW = B_{n-1} \qquad (n = 0, \cdots, N-1).$$

Then we have, by $(\mathrm{b})_N$ and (7.1.5),

$$\begin{aligned}s_N(Wg) &= \|XWg\|^2 + \sum_0^{N-1} \|B_nWg\|^2 = \|XWg\|^2 + \|B_{-1}g\|^2 + \sum_0^{N-2} \|B_mg\|^2 \\ &\le \|g\|^2 + \sum_0^{N-2} \|B_mg\|^2 = s_N(g) - \|B_{N-1}g\|^2,\end{aligned}$$

and hence, by $(\mathrm{a})_N$,

$$\|B_{N-1}g\|^2 = s_N(g) - s_N(Wg) \le \|g\|^2 - s_N(Wg) = \|Wg\|^2 - s_N(Wg) = \|D_NWg\|^2,$$

where $D_N = (I_{\mathfrak{G}} - X^*X - \Sigma_{n=0}^{N-1} B_n^*B_n)^{1/2}$.

Thus, again we conclude that there exists a contraction $C_N\colon \mathfrak{G} \to \mathfrak{K}$ such that $B_{N-1} = C_ND_NW$. If we define $B_N = C_ND_N$ we will have $B_NW = B_{N-1}$ and

$$\|B_Ng\|^2 = \|C_ND_Ng\|^2 \le \|D_Ng\|^2 = \|g\|^2 - s_N(g),$$

i.e., $(a)_{N+1}$ and $(b)_{N+1}$ hold.

Thus all B_n will be determined step by step, and (a), (b) will hold as limit cases of $(a)_N$ and $(b)_N$.

This concludes the proof.

2. Now we pass to the general "lifting theorem" for operators intertwining two contractions, say T and T', defined on the spaces $\mathfrak{H}$ and $\mathfrak{H}'$.

Let U_+ and U'_+ be the corresponding minimal isometric dilations defined on the spaces $\mathfrak{K}_+$ and $\mathfrak{K}'_+$, respectively. Let P_+ denote the orthogonal projection of $\mathfrak{K}_+$ onto $\mathfrak{H}$, and P'_+ that of $\mathfrak{K}'_+$ onto $\mathfrak{H}'$.

Theorem 2. *For every operator* $X: \mathfrak{H}' \to \mathfrak{H}$ *such that*

(a) $$TX = XT',$$

there exists an operator $Y: \mathfrak{K}'_+ \to \mathfrak{K}_+$ *satisfying*

(b) $$U_+Y = YU'_+,$$

(c′) $$X = P_+Y|\mathfrak{H}',$$

(c″) $$Y(\mathfrak{K}'_+ \ominus \mathfrak{H}') \subset \mathfrak{K}_+ \ominus \mathfrak{H},$$

and

(d) $$\|Y\| = \|X\|.$$

Conversely, every operator Y *satisfying* (b) *and* (c″) *gives rise, by the formula* (c′), *to an operator* X *satisfying* (a).

Proof. Let us first observe that the "lifting conditions" (c′), (c″) can be amalgamated in the single condition

(c) $$XP'_+ = P_+Y.$$

Suppose now that X satisfies (a). Multiplying by P_+ on the right and making use of the relation $T'P'_+ = P'_+U'_+$ (cf. (1.4.2)), we get

$$TX_0 = X_0U'_+, \quad \text{where } X_0 = XP'_+.$$

Applying Theorem 1 with $W = U'_+$ we conclude that there exists $Y: \mathfrak{K}'_+ \to \mathfrak{K}_+$ such that

$$U_+Y = YU'_+, \qquad X_0 = P_+Y, \qquad \|Y\| = \|X_0\| \quad (= \|X\|).$$

Hence we have $XP'_+ = X_0 = P_+Y$, i.e., (c).

Conversely consider any operator Y satisfying (b) and (c″). Using the relation (1.4.2) for T as well as for T', and the relation $P_+Y(I - P'_+) = 0$ equivalent to (c″), we conclude

$$TP_+Y|\mathfrak{H}' = P_+U_+Y|\mathfrak{H}' = P_+YU'_+|\mathfrak{H}' = P_+Y(P'_+ + I - P'_+)U'_+|\mathfrak{H}'$$
$$= P_+YP'_+U'_+|\mathfrak{H}' = P_+YT'P'_+|\mathfrak{H}' = P_+YT'.$$

Setting $X = P_+Y|\mathfrak{H}'$ we obtain (a) and the proof is done.

3. Let us denote the system of operators X satisfying condition (a) by $\mathfrak{I}(T', T)$, and the system of operators Y satisfying (b) and (c″) by $\mathfrak{I}^+(T', T)$. Then (c′) defines a map

$$\pi \colon \mathfrak{I}^+(T', T) \to \mathfrak{I}(T', T)$$

which is obviously linear and does not increase norm. By virtue of Theorem 2 this map is actually *onto*; moreover, for every $X \in \mathfrak{I}(T', T)$, there exists at least one $Y \in \pi^{-1}(X)$ such that $\|Y\| = \|X\|$.

The following property is an easy consequence of relation (c).

Multiplication property of π. If T_1, T_2, T_3 are any three contractions, and if $Y \in \mathfrak{I}^+(T_1, T_2)$ and $Z \in \mathfrak{I}^+(T_2, T_3)$, then

$$ZY \in \mathfrak{I}^+(T_1, T_3) \quad \text{and} \quad \pi_{13}(ZY) = \pi_{23}(Z)\pi_{12}(Y).$$

(The use of subscripts is self-explanatory.)

Also note that for one contraction T on $\mathfrak{H}$ and for the corresponding dilation space $\mathfrak{K}_+$, we obviously have

$$I_{\mathfrak{K}_+} \in \mathfrak{I}^+(T, T) \quad \text{and} \quad \pi(I_{\mathfrak{K}_+}) = I_{\mathfrak{H}}.$$

4. Returning to the case of two contractions, T and T', consider the Wold decompositions of the spaces $\mathfrak{K}_+$ and $\mathfrak{K}'_+$ generated by U_+ and U'_+, i.e.,

(7.4.1) $$\mathfrak{K}_+ = M_+(\mathfrak{L}_*) \oplus \mathfrak{R} \quad \text{and} \quad \mathfrak{K}'_+ = M_+(\mathfrak{L}'_*) \oplus \mathfrak{R}',$$

where

(7.4.2) $$\mathfrak{R} = \bigcap_0^\infty U_+^n\mathfrak{K}_+ \quad \text{and} \quad \mathfrak{R}' = \bigcap_0^\infty U_+'^n\mathfrak{K}'_+;$$

cf. (1.4.5) and (1.4.6).

For any operator Y satisfying $U_+Y = YU'_+$ we have

$$Y\mathfrak{R}' = \bigcap_0^\infty YU_+'^n\mathfrak{K}'_+ = \bigcap_0^\infty U_+^nY\mathfrak{K}'_+ \subset \bigcap_0^\infty U_+^n\mathfrak{K}_+ = \mathfrak{R};$$

therefore the decompositions (7.4.1) associate with such a Y there is a matrix representation of the form

$$Y = \begin{bmatrix} A & 0 \\ B & C \end{bmatrix}. \tag{7.4.3}$$

Setting $S_* = U_+|M_+(\mathfrak{L}_*)$, $R = U_+|\mathfrak{R}$, $S'_* = U'_+|M_+(\mathfrak{L}'_*)$, $R' = U'_+|\mathfrak{R}'$, we obviously have

$$A \in \mathfrak{I}(S'_*, S_*), \quad B \in \mathfrak{I}(S'_*, R), \quad C \in \mathfrak{I}(R', R). \tag{7.4.4}$$

Conversely, conditions (7.4.4) clearly imply $Y \in \mathfrak{I}(U'_+, U_+)$ for Y defined by (7.4.3).

Let us also use the decompositions

$$\mathfrak{K}_+ = \mathfrak{H} \oplus M_+(\mathfrak{L}) \quad \text{and} \quad \mathfrak{K}'_+ = \mathfrak{H}' \oplus M_+(\mathfrak{L}'); \tag{7.4.5}$$

cf. (1.4.4). The subspaces $M_+(\mathfrak{L})$ and $M_+(\mathfrak{L}')$ are invariant for (but not reducing) U_+ and U'_+; the restrictions

$$S = U_+|M_+(\mathfrak{L}) \quad \text{and} \quad S' = U'_+|M_+(\mathfrak{L}') \tag{7.4.6}$$

are unilateral shifts (so are S_* and S'_*). Consider the operators

$$\left.\begin{matrix} Q_+ \\ \\ Q_+^- \end{matrix}\right\} = \text{orthogonal projection of } M_+(\mathfrak{L}) \text{ into } \left\{\begin{matrix} M_+(\mathfrak{L}_*) \\ \\ \mathfrak{R} \end{matrix}\right.$$

and their analogues Q'_+ and Q'^-_+. We have

$$Q_+ \in \mathfrak{T}(S, S_*), \qquad Q_+^- \in \mathfrak{T}(S, R), \tag{7.4.7}$$

and analogous relations for Q'_+, Q'^-_+; cf. (2.3.3) and §4.2.

In order that an operator Y satisfying (7.4.3) and (7.4.4), and hence (b), also satisfy condition (c″), i.e., $YM_+(\mathfrak{L}') \subset M_+(\mathfrak{L})$, it is then necessary and sufficient that there exists an operator A_0: $M_+(\mathfrak{L}') \to M_+(\mathfrak{L})$ such that

$$AQ'_+ = Q_+A_0, \quad BQ'_+ + CQ'^-_+ = Q_+^-A_0 \quad (\text{on } M_+(\mathfrak{L}')). \tag{7.4.8}$$

(Then, necessarily, $A_0 = AQ'_+ + BQ'_+ + CQ'^-_+$.) Clearly (7.4.4) and (7.4.8) imply

$$A_0 \in \mathfrak{T}(S', S). \tag{7.4.9}$$

Thus *the general form of an operator* Y *satisfying conditions* (b) *and* (c″) *of Theorem* 2 *is a matrix* (7.4.3) *satisfying conditions* (7.4.4), (7.4.8) *and* (7.4.9).

The map π: $\mathfrak{I}^+(T', T) \to \mathfrak{I}(T', T)$ is in general not one-to-one. Its kernel consists of those $Y \in \mathfrak{I}^+(T', T)$ for which $Y\mathfrak{H}' \subset \mathfrak{H}^\perp$ $(= M_+(\mathfrak{L}))$. Since we have $YM_+(\mathfrak{L}') \subset M_+(\mathfrak{L})$ by (c″), condition $\pi(Y) = 0$ is equivalent to condition

$$Y\mathfrak{K}'_+ \subset M_+(\mathfrak{L}). \tag{7.4.10}$$

Now (7.4.10) implies first that

$$Y\mathfrak{R}' = \bigcap_0^\infty YU_+'^n\mathfrak{R}_+' = \bigcap_0^\infty U_+^n Y\mathfrak{R}_+' \subset \bigcap_0^\infty S^n M_+(\mathfrak{L}) = \{0\}$$

(because S is a unilateral shift); hence $C = 0$. It also implies that, in particular, $YM_+(\mathfrak{L}'_*) \subset M_+(\mathfrak{L})$; hence we deduce that the operator $D = Y|M_+(\mathfrak{L}'_*)$ belongs to $\mathfrak{I}(S'_*, S)$ and that for $x \in M_+(\mathfrak{L}'_*)$ we have

$$\begin{bmatrix} A & 0 \\ B & C \end{bmatrix}\begin{bmatrix} x \\ 0 \end{bmatrix} = Yx = Dx = \begin{bmatrix} Q_+Dx \\ Q_+^-Dx \end{bmatrix}.$$

Hence, $A = Q_+D$, $B = Q_+^-D$.

Conversely, if D is any operator satisfying

(7.4.11) $$D \in \mathfrak{I}(S'_*, S)$$

then the operators defined by

(7.4.12) $$A = Q_+D, \quad B = Q_+^-D, \quad C = 0, \quad A_0 = DQ'_+$$

clearly fulfill conditions (7.4.8), and hence the corresponding operator $Y = \begin{bmatrix} A & 0 \\ B & C \end{bmatrix}$ belongs to $\mathfrak{I}^+(T', T)$. Moreover, we have

$$Y\begin{bmatrix} x \\ y \end{bmatrix} = \begin{bmatrix} Q_+Dx \\ Q_+^-Dx \end{bmatrix} = Dx \in M_+(\mathfrak{L}) \quad \text{for any} \quad \begin{bmatrix} x \\ y \end{bmatrix} \in \mathfrak{R}'_+,$$

i.e., $Y\mathfrak{R}'_+ \subset M_+(\mathfrak{L})$ and therefore $\pi(Y) = 0$.

Thus *the general form of an operator* $Y \in \mathfrak{I}^+(T', T)$ *satisfying* $\pi(Y) = 0$ *is*

$$Y = \begin{bmatrix} Q_+D & 0 \\ Q_+^-D & 0 \end{bmatrix} \quad \textit{with } D \in \mathfrak{I}(S'_*, S).$$

The above structural properties of the lifting yield in particular an approach to the problem of finding conditions under which two contractions are *similar*. We do not go into the details of such applications.

5. Consider the case that T and T' are cnu and given by their functional models, i.e., $T = S(\Theta)$, $T' = S(\Theta')$, where $\{\mathfrak{A}, \mathfrak{A}_*, \Theta(\lambda)\}$ and $\{\mathfrak{A}', \mathfrak{A}'_*, \Theta'(\lambda)\}$ are purely contractive analytic functions (see §3). Then,

$$\mathfrak{R}_+ = H^2(\mathfrak{A}_*) \oplus \overline{\Delta L^2(\mathfrak{A})},$$

$$U_+ = \text{multiplication by } \chi,$$

$$\mathfrak{H} = \mathfrak{K}_+ \ominus \{\Theta u \oplus \Delta u:\ u \in H^2(\mathfrak{A})\},$$

$$T = P_+ U_+ | \mathfrak{H};$$

$$\mathfrak{L}_* = \{a_* \oplus 0:\ a_* \in \mathfrak{A}_*\},$$

$$\mathfrak{L} = \{\Theta a \oplus \Delta a:\ a \in \mathfrak{A}\},$$

$$M_+(\mathfrak{L}_*) = \{u_* \oplus 0:\ u_* \in H^2(\mathfrak{A}_*)\},$$

$$M_+(\mathfrak{L}) = \{\Theta u \oplus \Delta u:\ u \in H^2(\mathfrak{A})\},$$

and therefore

$$\left.\begin{aligned} Q_+(\Theta u \oplus \Delta u) &= \Theta u \oplus 0 \\ Q_+^-(\Theta u \oplus \Delta u) &= 0 \oplus \Delta u \end{aligned}\right\} \quad (u \in H^2(\mathfrak{A}));$$

similarly for T.

Now we use the fact that if $\mathfrak{S}$ and $\mathfrak{F}$ are (separable) Hilbert spaces and $\{\mathfrak{S}, \mathfrak{S}, \Omega(e^{it})\}$, $\{\mathfrak{S}', \mathfrak{S}', \Omega'(e^{it})\}$ are bounded measurable functions, then those operators

(a) $\Phi:\ H^2(\mathfrak{S}') \to H^2(\mathfrak{S})$,

(b) $\Phi:\ H^2(\mathfrak{S}') \to \overline{\Omega L^2(\mathfrak{S})}$,

(c) $\Phi:\ \overline{\Omega' L^2(\mathfrak{S}')} \to \overline{\Omega L^2(\mathfrak{S})}$,

which commute with multiplication by χ, are themselves multiplication (on the left) by operator valued, bounded functions $\{\mathfrak{S}', \mathfrak{S}, \Phi(e^{it})\}$, which are, respectively,

(a) analytic,

(b) measurable with $\Phi(e^{it})\mathfrak{S}' \subset \overline{\Omega(e^{it})\mathfrak{S}}$ a.e.,

(c) measurable with $\Phi(e^{it})\Omega'(e^{it})\mathfrak{S}' \subset \overline{\Omega(e^{it})\mathfrak{S}}$ a.e.

We conclude

Theorem 3. *The operators $Y \in \mathfrak{I}^+(T', T)$ are multiplication on $\mathfrak{K}_+$ by matrix functions of the form*

$$Y(e^{it}) = \begin{bmatrix} A(e^{it}) & 0 \\ B(e^{it}) & C(e^{it}) \end{bmatrix}$$

where $\{\mathfrak{A}'_*, \mathfrak{A}_*, A(\lambda)\}$ is a bounded analytic function while $\{\mathfrak{A}'_*, \mathfrak{A}_*, B(e^{it})\}$ and $\{\mathfrak{A}'_*, \mathfrak{A}_*, C(e^{it})\}$ are bounded measurable functions, satisfying the conditions

$$\left.\begin{aligned} B(e^{it})\mathfrak{A}'_* &\subset \overline{\Delta(e^{it})\mathfrak{A}_*}, \\ C(e^{it})\Delta'(e^{it})\mathfrak{A}'_* &\subset \overline{\Delta(e^{it})\mathfrak{A}_*}, \end{aligned}\right\} \quad a.e.$$

and

$$A(e^{it})\Theta'(e^{it}) = \Theta(e^{it})A_0(e^{it}),$$

$$\text{a.e.},$$

$$B(e^{it})\Theta'(e^{it}) + C(e^{it})\Delta'(e^{it}) = \Delta(e^{it})A_0(e^{it}),$$

where $\{\mathfrak{A}', \mathfrak{A}, A_0(\lambda)\}$ *is another bounded analytic function.*

For any $X \in \mathcal{I}(T', T)$ *there is a* Y *of the above type such that* $X = P_+ Y|\mathfrak{H}'$ *and* $\|Y\|_\infty = \|X\|$.

Note that if both T and T' are of class $C_{\cdot 0}$, i.e., if Θ and Θ' are *inner* functions, then the second components disappear so that, in this case, Theorem 3 takes the following form:

Theorem 3′. *If* $T = S(\Theta)$, $T' = S(\Theta')$ *with inner* $\{\mathfrak{A}, \mathfrak{A}_*, \Theta(\lambda)\}$, $\{\mathfrak{A}', \mathfrak{A}'_*, \Theta'(\lambda)\}$ *then the operators* $Y \in \mathcal{I}^+(T', T)$ *are multiplication on* $H^2(\mathfrak{A}'_*)$ *by a bounded analytic function* $\{\mathfrak{A}'_*, \mathfrak{A}_*, Y(\lambda)\}$ *satisfying*

(7.5.1) $$Y(\lambda)\Theta'(\lambda) = \Theta(\lambda)Y_0(\lambda),$$

where $\{\mathfrak{A}', \mathfrak{A}, Y_0(\lambda)\}$ *is again a bounded analytic function.*[4] *For any* $X \in \mathcal{I}(T', T)$ *there exists a function* $Y(\lambda)$ *of this type such that*

(7.5.2) $$Xu = P_+ Yu \quad \text{for } u \in \mathfrak{H}' \ (= H^2(\mathfrak{A}'_*) \ominus \Theta' H^2(\mathfrak{A}')), \quad \text{and}$$
$$\|Y\|_\infty = \|X\|.$$

The case where $\Theta(\lambda)$ and $\Theta'(\lambda)$ are *scalar* valued inner functions and *equal* is particularly simple: then $Y(\lambda)$ is also scalar valued and (7.5.1) is obviously fulfilled. This particular case of the theorem was proved by another method by D. Sarason [7]: his result suggested the above investigations.

These theorems have found numerous applications in operator theory and in analysis. A minor application will be given in §10.

---

[4] This requirement is equivalent to the following one:

(7.5.1)′ $$Y' H^2(\mathfrak{A}') \subset \Theta H^2(\mathfrak{A}).$$

8. Functional Calculus for a Contraction

1. If T is a cnu contraction on $\mathfrak{H}$, then we can define, for any function

(8.1.1) $$u(\lambda) = \sum_0^\infty a_n \lambda^n \in H^\infty,$$

i.e., analytic and bounded in the open unit disc ($|\lambda| < 1$), an operator $u(T)$ on $\mathfrak{H}$ as the "Abel sum" of the operator series $\Sigma_0^\infty a_n T^n$, i.e., by

(8.1.2) $$u(T) = \lim_{r \to 1-0} \sum_0^\infty r^n a_n T^n.$$

As $\limsup_{n \to \infty} |a_n|^{1/n} \leq 1$, the series on the right-hand side of (8.1.2) converges for fixed r ($0 \leq r < 1$) in operator norm. Set

(8.1.3) $$u_r(T) = \sum_0^\infty r^n a_n T^n \quad \text{and} \quad u_r(U) = \sum_0^\infty r^n a_n U^n,$$

where U is the minimal unitary dilation of T, on the space $\mathfrak{K}$. Since we have

(8.1.4) $$u_r(T) = P_{\mathfrak{H}} u_r(U)|\mathfrak{H},$$

$u_r(T)$ strongly converges as $r \to 1-0$ if $u_r(U)$ does. Now if $E(\sigma)$ is the spectral measure function associated with the unitary operator u then we have, for $0 < r, \rho < 1$ and $k \in \mathfrak{K}$,

(8.1.5) $$\|[u_r(U) - u_\rho(U)]k\|^2 = \int_0^{2\pi} |u(re^{it}) - u(\rho e^{it})|^2 d\|E_t k\|^2.$$

The integrand is bounded by $4\|u\|_\infty^2$ and as $r, \rho \to 1-0$ it converges to 0, by Fatou's theorem, a.e. with respect to Lebesgue measure.

Now, the operator U is either itself a bilateral shift or the "skew sum" of two bilateral shifts in the sense of the formula (2.2.3); as the spectral measure of a bilateral shift is absolutely continuous with respect to Lebesgue measure, [5] we conclude

---

[5] The spectral measure $F(\sigma)$ of a bilateral shift of multiplicity d is unitarily equivalent to the operator of multiplication by the indicator function of the set σ on the orthogonal sum of d copies of the space L^2 (with respect to normalized Lebesgue measure on the unit circle).

that *the spectral measure* $E(\sigma)$ *of* U *is absolutely continuous* too.

Thus (8.1.5) implies $\|u_r(U)k - u_\rho(U)k\| \to 0$ as $r, \rho \to 1-0$, for every $k \in \mathfrak{K}$, and this concludes the proof of the existence of the strong limit in the definition (8.1.2).

[Let us observe that reference to Fatou's theorem can be avoided by only using that $u(re^{it}) - u(\rho e^{it}) = \sum_1^\infty (r^n - \rho^n)a_n e^{int}$ tends to 0 in L^2 as $r, \rho \to 1$, and by splitting $f(t) = (d/dt)\|E_t k\|^2$ in the sum of a bounded $f_1(t)$ and an $f_2(t)$ with $\int_0^{2\pi}|f_2(t)|dt < \epsilon$.]

It is obvious that $u \mapsto u(T)$ is an algebra homomorphism of H^∞ into the algebra of operators on $\mathfrak{H}$, with the further properties:

(α) $1 \mapsto I$, $\lambda \mapsto T$.

(β) $u(T)^* = u^\sim(T^*)$.

(γ) $\|u(T)\| \leq \|u\|_\infty$ (*von Neumann inequality, cf.* (6.4.1)).

As a consequence of this inequality, if $\{u_n\}$ is a sequence in H^∞ then

(δ) $\|u_n\|_\infty \to 0$ *implies* $\|u_n(T)\| \to 0$.

Other continuity properties:

(ϵ) *If* $\|u_n\|_\infty \leq C$ *and* $u_n(e^{it}) \to 0$ *a.e., then* $u_n(T) \to 0$ *strongly.*

(η) *If* $\|u_n\|_\infty \leq C$ *and* $u_n(\lambda) \to 0$ *at every point* λ *in the open unit disc, then* $u_n(T) \to 0$ *weakly.*

Property (ϵ) follows from the absolute continuity of the spectral measure $E(\sigma)$ and from Lebesgue's theorem, similar to the proof of the existence of the strong limit in (8.1.2). As for (η), observe first that, for any integer ν,

$$\int_0^{2\pi} u_n(e^{it})e^{i\nu t}dt = \frac{1}{i}\oint_{|\lambda|=1} u_n(\lambda)\lambda^{\nu-1}d\lambda$$
$$= \frac{1}{i}\oint_{|\lambda|=1/2} u_n(\lambda)\lambda^{\nu-1}d\lambda \to 0;$$

hence, for any function $f(t) \in L^1$, arguing with the Fejér means of the Fourier series of $f(t)$,

$$\int_0^{2\pi} u_n(e^{it})f(t)dt \to 0.$$

Again using the absolute continuity of $E(\sigma)$ we conclude

$$(u_n(T)h, g) = \int_0^{2\pi} u_n(e^{it})d(E_t f, g) \to 0 \quad \text{for } f, g \in \mathfrak{H} \text{ and } n \to \infty.$$

2. As a further property of this functional calculus let us prove the following:

(ι) *If* $u \in H^\infty$ *and* $|u(\lambda)| < 1$ *for* $|\lambda| < 1$, *then* $T' = u(T)$ *is also a cnu contraction, and we have*

(8.2.1) $$v(T') = v \circ u(T) \quad \textit{for every } v \in H^\infty.$$

Proof. Set $u_0(\lambda) = [u(\lambda) - u(0)][1 - \overline{u(0)}u(\lambda)]^{-1}$; then $u_0 \in H^\infty$ and $|u_0(\lambda)| < 1$ for $|\lambda| < 1$. As $u_0(0) = 0$ we have, by the Schwartz lemma, $u_0(\lambda) = \lambda \cdot u_1(\lambda)$, where $|u_1(\lambda)| \leq 1$ for $|\lambda| < 1$. Hence $\|u_0(T)\| \leq 1$, $\|u_1(T)\| \leq 1$, and, for every $h \in \mathfrak{H}$ and integer $n \geq 0$,

$$\|u_0(T)^n h\| = \|u_1(T)^n T^n h\| \leq \|T^n h\|,$$

$$\|u_0(T)^{*n} h\| = \|u_1(T)^{*n} T^{*n} h\| \leq \|T^{*n} h\|.$$

As T is cnu, we have, for every $h \neq 0$, $\inf_n \{\|T^n h\|, \|T^{*n}h\|\} < \|h\|$, and this implies that $T_0 = u_0(T)$ is also a cnu contraction. This property is inherited by the operator

$$T' = [T_1 + u(0)I][I + \overline{u(0)}T_1]^{-1}$$

because, from this relation, it is easy to infer that T' and T_1 have the same space of unitarity.

As T' is a cnu contraction, $v(T')$ makes sense for every $v \in H^\infty$, $v(\lambda) = \Sigma_0^\infty a_n \lambda^n$. The relation $p(T') = p \circ u(T)$ is immediate for any polynomial p; hence, in particular, for the Fejér sums

$$v_n(\lambda) = \sum_0^n (1 - k/n) a_k \lambda^k;$$

thus

$$v_n(T') = v_n \circ u(T). \tag{8.2.2}$$

Now we have $\|v_n\|_\infty \leq \|v\|_\infty$, $\|v_n \circ u\|_\infty \leq \|v\|_\infty$, and $v_n(\lambda) \to v(\lambda)$, $v_n \circ u(\lambda) \to v \circ u(\lambda)$ if $|\lambda| < 1$ so (8.2.2) implies (8.2.1) as $n \to \infty$, by property (η) of the functional calculus, when applied to $v_n - v$ and $v_n \circ u - v \circ u$.

3. Let us recall the following facts for functions of Hardy class H^1 (on the unit circle):

For any given function $k \in L^1$ such that $k \geq 0$ and $\log k \in L^1$ the function

$$v(\lambda) = \exp\left[\frac{1}{2\pi}\int_0^{2\pi} \frac{e^{it} + \lambda}{e^{it} - \lambda} \log k(t)dt\right] \qquad (|\lambda| < 1), \tag{8.3.1}$$

or rather its boundary value on the unit circle, belongs to H^1 and is called an *outer* function; we have

$$|v(e^{it})| = k(t) \quad \text{a.e.}$$

On the other hand, for every nonzero function $u \in H^1$ we have $\log|u(e^{it})| \in L^1$ so that the outer function v corresponding to $k(t) = |u(e^{it})|$ exists: this function v is called the *outer factor* of u and will be also denoted by u_e. The quotient $u_i = u/u_e$ is an *inner* function (i.e., $u_i \in H^\infty$ and $|u_i(e^{it})| = 1$ a.e.).

Inner functions have a canonical factorization into a Blaschke product and a function of the form

$$\exp\left[-\int_0^{2\pi} \frac{e^{it}+\lambda}{e^{it}-\lambda}\, d\mu_t\right], \tag{8.3.2}$$

where $\mu(\sigma)$ is a nonnegative, finite Borel measure, singular with respect to Lebesgue measure.

One of the facts which can be deduced by using these factorizations is that, for every family $\{u_\alpha\}$ of inner functions, there exists a *greatest common* (inner) *divisor*

$$v = \bigwedge_\alpha u_\alpha.$$

We need the following:

Lemma. *Let* $\{u_\alpha = u_{\alpha i} \cdot u_{\alpha e}\}$ *be a family of nonzero functions of class* H^1 *and set* $v = \bigwedge_\alpha u_{\alpha i}$. *Suppose* $f \in L^1$ *is such that* $F_\alpha := u_\alpha f \in H^1$ *for all* α. *Then*

$$vf \in H^1.$$

Proof. Fix an α, say α_0, and set $u = u_{\alpha_0}$ and $F = F_{\alpha_0}$. Then $F_\alpha u = F u_\alpha$ $(= uu_\alpha f)$ for every α, and hence

$$\log|F_\alpha| + \log|u| = \log|F| + \log|u_\alpha|;$$

thus by (8.3.1), and by division,

$$F_{\alpha e}u_e = F_e u_{\alpha e} \quad \text{and} \quad F_{\alpha i}u_i = F_i u_{\alpha i}.$$

Set $u'_{\alpha i} = u_{\alpha i}/v$ and in particular $u'_i = u_i/v$; then

$$F_{\alpha i}u'_i = F_i u'_{\alpha i}.$$

As the family $\{u'_{\alpha i}\}$ is prime, we deduce that u'_i is a divisor of F_i, i.e., F_i/u'_i is an inner function. Now we have

$$vf = \frac{u_i}{u'_i} f = \frac{uf}{u'_i u_e} = \frac{F}{u'_i u_e} = \frac{F_i}{u'_i}\frac{F_e}{u_e} \in H^1$$

because F_i/u'_i is inner and F_e/u_e is in H^1; indeed we have

$$\frac{F_e(\lambda)}{u_e(\lambda)} = \exp\left[\frac{1}{2\pi}\int_0^{2\pi} \frac{e^{it}+\lambda}{e^{it}-\lambda}\log|f(t)|\,dt\right]$$

where $\log|f(t)| = \log|F(e^{it})| - \log|u(e^{it})| \in H^1$ and, by assumption, $f \in L^1$.

Noting that $H^\infty \subset H^1$ we can apply this lemma to prove the following:

Theorem. *Let T be a cnu contraction on the space $\mathfrak{H}$, $\{u_\alpha\}$ a family of nonzero functions of class H^∞, and h an element of $\mathfrak{H}$ such that*

$$u_\alpha(T)h = 0 \quad \text{for all } \alpha. \tag{8.3.3}$$

Then we also have

$$v(T)h = 0, \quad \text{where } v = \bigwedge_\alpha u_{\alpha i}. \tag{8.3.4}$$

Proof. Observe that (8.3.3) implies, for every $g \in \mathfrak{H}$ and every integer $m \geq 0$,

$$0 = (T^m u_\alpha(T)h, g) = \int_0^{2\pi} e^{imt} u_\alpha(e^{it}) \frac{d}{dt}(E_t h, g)dt;$$

setting

$$f(t) = e^{-it}\frac{d}{dt}(E_t h, g)$$

we have therefore $u_\alpha f \in H^1$. Since moreover $f \in L^1$ we conclude that $vf \in H^1$, by the lemma. Thus,

$$(v(T)h, g) = \int_0^{2\pi} v(e^{it})\frac{d}{dt}(E_t h, g)dt = \int_0^{2\pi} vf \cdot e^{it}dt = 0.$$

As g is arbitrary in $\mathfrak{H}$ this gives (8.3.4).

Corollary 1. *If u is outer, $u \in H^\infty$, then $u(T)$ is a quasi-affinity.*[6]

Proof. If $u(T)h = 0$ for an $h \in \mathfrak{H}$ then $v(T)h = 0$ for $v = u_i$ = const of modulus 1, by the theorem. Thus $h = 0$, i.e., $u(T)$ has zero kernel. If $u(T)^*h = 0$ for an $h \in \mathfrak{H}$ then as $u(T)^* = u^\sim(T^*)$, and as $u^\sim$ is also outer, we have, by the same reason, $h = 0$, i.e., $u(T)$ has dense range in $\mathfrak{H}$.

Let us introduce the *class* C_0 of operators: it consists of all cnu contractions T such that $u(T) = 0$ for some nonzero $u \in H^\infty$.

Corollary 2. *For every $T \in C_0$ there exists an inner function m for which $m(T) = 0$ and which is a divisor (in H^∞) of every other function $u \in H^\infty$ such that $u(T) = 0$.*

Proof. Take $m = \bigwedge u_i$ for all such u.

---

[6] An operator X on $\mathfrak{H}$, or more generally, from $\mathfrak{H}$ into $\mathfrak{H}'$, is called a *quasi-affinity* if it has zero kernel in $\mathfrak{H}$ and dense range in $\mathfrak{H}'$.

If we disregard of constant factors of modulus 1 then this is the only function with the required properties: it is called the *minimal function* of T, and denoted by m_T.

From property (β) of our functional calculus it follows that $T^* \in C_0$ whenever $T \in C_0$, and that $m_{T^*} = (m_T)^\sim$.

Corollary 3. *If* $T \in C_0$ *and if* $u \in H^\infty$ *is such that* $u_i \wedge m_T = 1$, *then* $u(T)$ *is a quasi-affinity.*

Proof. This proof is analogous to the proof of Corollary 1, with the difference that here we also have $m_T(T)h = 0$ and $\tilde{m}_T(T^*)h = 0$ for all h because $m_T(T) = 0$ and $\tilde{m}_T(T^*) = m_T(T)^* = 0$.

4. The functional calculus for a cnu contraction T on $\mathfrak{H}$ can be extended to some *unbounded* functions in the following way:

Let N_T denote the *class* of functions which are of the form

$$\phi = u/v, \tag{8.4.1}$$

where $u, v \in H^\infty$ and $v(T)$ is a quasi-affinity. For such a function ϕ define $\phi(T) = v(T)^{-1}u(T)$.

[Note that $v(T)^{-1}$, and hence $\phi(T)$, are in general not bounded operators.]

This definition of $\phi(T)$ does not depend on the particular choice of the representation (8.4.1). For, if $\phi = u/v = u'/v'$, then the relation $uv' = vu'$ implies $u(T)v'(T) = v(T)u'(T)$, and hence

$$\begin{aligned} v(T)^{-1}u(T) &= v(T)^{-1}v'(T)^{-1}v'(T)u(T) = v'(T)^{-1}v(T)^{-1}u(T)v'(T) \\ &= v'(T)^{-1}v(T)^{-1}v(T)u'(T) = v'(T)^{-1}u'(T). \end{aligned}$$

Thus the definition of $\phi(T)$ is *unique.*

Some further facts whose proof is more or less routine:

(a) The class N_T is an *algebra.*

(b) For every $\phi \in N_T$, $\phi(T)$ is *closed* and with *dense domain* in $\mathfrak{H}$.

(c) For $\phi \in N_T$ and every (bounded) operator A commuting with T, $\phi(T)A \supset A\phi(T)$.

(d) $(c\phi)(T) = c \cdot \phi(T)$ if $\phi \in N_T$ and $c \neq 0$,
$(\phi_1 + \phi_2)(T) \supset \phi_1(T) + \phi_2(T)$,
$(\phi_1\phi_2)(T) \supset \phi_1(T)\phi_2(T)$
for $\phi_1, \phi_2 \in N_T$, with "$=$" in place of "$\supset$" in particular if $\phi_2 \in H^\infty$.

(e) $\phi^\sim(T^*) \supset \phi(T)^*$.

(f) If T is normal, $T = \int \lambda dE_\lambda$, then $\phi(T) = \int \phi(\lambda) dE_\lambda$ for $\phi \in N_T$.

5. The functional calculus extends to contractions T of general type, on the basis of the canonical decomposition $T = T_0 \oplus W$ in an orthogonal sum of a cnu contraction T_0 and a unitary operator W. In order that the definition (8.1.2) make sense in this case too, one has to restrict the class H^∞ of functions to the class, denoted by H_T^∞, of functions $u \in H^\infty$ for which $u(e^{it})$ (the radial limit) exists *a.e. with respect to the spectral measure of the unitary part* W. Most of the properties of the calculus observed above carry over to this case.

9. Operators of Class C_0 and the Jordan Model

1. Operators of class C_0 were defined in §8.4 as those cnu contractions T for which there exists a nonzero function $u \in H^\infty$ such that $u(T) = 0$. Among these functions there is a minimal one, i.e., which is a divisor in H^∞ of all the others and which is, moreover, an inner function. This minimal function, denoted by m_T, is determined up to a constant factor of modulus 1.

Operators of class C_0 have interesting properties. Although they are mostly "genuinely infinite-dimensional," they behave in many respects in close analogy to operators on finite-dimensional space.

We know from §2.1 that for every contraction T on the space $\mathfrak{H}$ and for its (minimal) unitary dilation U on the space $\mathfrak{K}$ the (strong) limit

(9.1.1) $$L = \lim_{n\to\infty} U^{-n}T^n$$

exists: L is an operator from $\mathfrak{H}$ into $\mathfrak{K}$. Clearly, $LT^m = U^mL$ $(m = 0, 1, 2, \cdots)$.

If $T \in C_0$ then this implies

(9.1.2) $$0 = L \cdot m_T(T) = m_T(U) \cdot L,$$

where $m_T(U)$ is given through the spectral integral $\int_0^{2\pi} m_T(e^{it})dE_t$. As $|m_T(e^{it})| = 1$ a.e. on the unit circle, and as the spectral measure $E(\sigma)$ of U is absolutely continuous with respect to Lebesgue measure, $m_T(U)$ is a unitary operator, and hence (9.1.2) implies $L = 0$, and, by (9.1.1), $T^n \to 0$. Because T^* belongs to C_0 along with T, we also have $T^{*n} \to 0$. We can express these results by the inclusion formula

(9.1.3) $$C_0 \subset C_{00}.$$

The converse inclusion is not true; indeed, there are even "strict" contractions T (i.e., for which $\|T\| < 1$) not belonging to C_0.

This follows from the fact that the spectrum of an operator $T \in C_0$ *inside the unit circle* is a discrete set consisting of the zeros of the minimal function $m_T(\lambda)$; each zero of $m_T(\lambda)$ is an eigenvalue of T.

To prove this one observes that the function

$$n(\lambda_0;\ \lambda) = (m_T(\lambda) - m_T(\lambda_0))/(\lambda - \lambda_0)$$

belongs to H^∞; thus $n(\lambda_0; T)$ makes sense as a (bounded) operator and satisfies

$$n(\lambda_0; T) \cdot (T - \lambda_0 I) = (T - \lambda_0 I) \cdot n(\lambda_0; T) = - m_T(\lambda_0) I.$$

Thus if $m_T(\lambda_0) \neq 0$ then $(T - \lambda_0 I)^{-1} = -(1/m_T(\lambda_0)) n(\lambda_0; T)$, and hence λ_0 belongs to the resolvent set of T. If $m_T(\lambda_0) = 0$ then $n(\lambda_0; \lambda) = m_T(\lambda)/(\lambda - \lambda_0)$ is not a multiple of $m_T(\lambda)$, and hence $n(\lambda_0; T) \neq 0$; and every nonzero vector in the range of $n(\lambda_0; T)$ is an eigenvector of T.

Let us add (without going into the proof) that the spectrum of T *on the unit circle* consists of the points through which the function $m_T(\lambda)$ cannot be extended analytically into the exterior.

2. From §3.3 and from the relation (4.1.4) between the characteristic functions of T and T^* it follows that the functional model of a contraction $T \in C_{00}$ is the operator $S(\Theta)$ associated with a purely contractive analytic function $\{\mathfrak{A}, \mathfrak{A}_*, \Theta(\lambda)\}$ which is inner together with $\{\mathfrak{A}_*, \mathfrak{A}, \Theta^\sim(\lambda)\}$, i.e., for which the values $\Theta(e^{it})$ are unitaries a.e.; thus there is no restriction to assume that $\mathfrak{A} = \mathfrak{A}_*$. Recall that, in this case, $S(\Theta)$ is defined on the space $\mathfrak{H}(\Theta) = H^2(\mathfrak{A}) \ominus \Theta H^2(\mathfrak{A})$ by

$$S(\Theta)h = P_{\mathfrak{H}(\Theta)}(\chi h) \qquad (h \in \mathfrak{H}(\Theta)).$$

We have then

$$m(S(\Theta))h = P_{\mathfrak{H}(\Theta)}(mh) \qquad (h \in \mathfrak{H}(\Theta))$$

for any function $m \in H^\infty$. As we have $m \cdot \Theta H^2(A) \subset \Theta H^2(A)$, it follows that $m(S(\Theta)) = 0$ if and only if

$$m \cdot H^2(A) \subset \Theta H^2(A). \tag{9.2.1}$$

We can express this relation by saying that the scalar valued function $m \in H^\infty$ is a "multiple" of the operator valued function Θ.

If $\Theta(\lambda)$ itself is scalar valued, i.e., $\dim \mathfrak{A} = 1$ and $\Theta(\lambda) = a(\lambda) \in H^\infty$, then (9.2.1) is fulfilled by $m = a$, and every other $m \in H^\infty$ fulfilling (9.2.1) is a multiple of a in H^∞. Thus we have

Theorem 1. *For every given nonconstant scalar valued inner function m, the operator $S(m)$ defined on the space $\mathfrak{H}(m) = H^2 \ominus mH^2$ by*

$$S(m)h = P_{\mathfrak{H}(m)}(\chi h) \qquad (h \in \mathfrak{H}(m))$$

is of class C_0 and has the minimal function $m_T = m$.

These operators $S(m)$ are, in many respects the simplest type of operator of class C_0. They are *cyclic*, i.e., there exists $h \in \mathfrak{H}(m)$ such that $h, S(m)h, S(m)^2 h, \ldots$ span the space $\mathfrak{H}(m)$. Indeed, so is the function

(9.2.2) $$h(\lambda) = 1 - \overline{m(0)}m(\lambda).$$

The fact $h \in \mathfrak{H}(m)$ follows from the relation $\overline{m}h = \overline{m} - \overline{m(0)} \perp H^2$. To prove that $S(m)^n h$ $(n = 0, 1, \cdots)$ span $\mathfrak{H}(m)$, consider a $v \in \mathfrak{H}(m)$ orthogonal to these vectors. Then, for $n = 0, 1, \cdots,$

$$0 = (v, P_{\mathfrak{H}(m)}(\chi^n h)) = (v, \chi^n) - m(0)(v, \chi^n m) = (v, \chi^n)$$

because $\chi^n m \in mH^2$; as $v \in H^2$ this implies $v = 0$.

Let us also note that $S(m)^*$ is unitarily equivalent to $S(m^\sim)$. Namely, we have

$$\Psi S(m)^* = S(m^\sim)\Psi,$$

where Ψ: $\mathfrak{H}(m) \to \mathfrak{H}(m^\sim)$ is the unitary operator defined by

$$(\Psi u)(e^{it}) = e^{-it} \cdot m^\sim(e^{it})u(e^{-it}) \qquad (u \in \mathfrak{H}(m)).$$

(Exercise!) Hence it follows in particular that the operator $S(m)^*$ is also *cyclic*.

3. It is convenient to use the following notation. For any operator T on a space $\mathfrak{H}$ define the *multiplicity* μ_T as the least cardinal number of a set of vectors in $\mathfrak{H}$ which, together with their transforms by $T, T^2, \cdots, T^n, \cdots$ span the space $\mathfrak{H}$.

Thus in particular $\mu_T = 1$ means that T is cyclic, i.e. has a vector v such that $v, Tv, \cdots, T^n v, \cdots$ span the space.

For two operators, say T_1 on the space $\mathfrak{H}_1$ and T_2 on the space $\mathfrak{H}_2$, we call T_2 a *quasi-affine transform* of T_1 if there exists a quasi-affinity A: $\mathfrak{H}_2 \to \mathfrak{H}_1$ such that

(9.3.1) $$T_1 A = A T_2,$$

and then we denote $T_1 \succ T_2$ or $T_2 \prec T_1$.

As the adjoint of a quasi-affinity is also a quasi-affinity, $T_1 \succ T_2$ implies $T_2^* \succ T_1^*$.

In case $T_1 \succ T_2$ and $T_1 \prec T_2$ simultaneously hold, we call T_1 and T_2 *quasi-similar.*

From (9.3.1) it follows $T_1^n A = A T_2^n$ for $n = 0, 1, \cdots$; hence we deduce that $T_1 \succ T_2$ implies $\mu_{T_1} \le \mu_{T_2}$.

For every operator T on a finite-dimensional space we have $\mu_T = \mu_{T^*}$. For operators on an infinite-dimensional space this is not always the case: e.g., if S is a simple unilateral shift, say $S\langle \xi_0, \xi_1, \cdots\rangle = \langle 0, \xi_0, \xi_1, \cdots\rangle$ on l^2, then for the operator $T = S \oplus S$ we have $\mu_T = 2$ and $\mu_{T^*} = 1$ (D. E. Sarason).

The operators of class C_0 are well behaved in this respect, for the following theorems hold:

Theorem A. *For any* T *of class* C_0 *we have* $\mu_T = \mu_{T^*}$. *If* $\mu_T < \infty$ *then* T *is* *quasi-similar to a uniquely determined operator of the form*

$$S(m_1) \oplus S(m_2) \oplus \cdots \oplus S(m_K), \tag{9.3.2}$$

where $m_1, m_2, \cdots, m_K$ *are (scalar valued) nonconstant inner functions each of which is a divisor (in* H^∞*) of its predecessor; here we have* $K = \mu_T$.

This operator (9.3.2) is called the "Jordan model" of T because it resembles in some respects the usual Jordan normal form of a finite matrix.

For the particular case $\mu_T = 1$ (i.e., for cyclic or "multiplicity-free" operators) we have the following, more complete result:

Theorem B. *For an operator* T *of class* C_0 *the following conditions are equivalent:*

(i) $\mu_T = 1$.
(i$_*$) $\mu_{T^*} = 1$.
(ii) T *is quasi-similar to* $S(m)$, $m = m_T$.

(i), (i$_*$), (ii): special case of Theorem A

(iii) *For every inner function* m *dividing* m_T *there is a unique invariant subspace* $\mathfrak{L}$ *for* T *such that the minimal function of* $T|\mathfrak{L}$ *equals* m*; in fact,* $\mathfrak{L} = \ker m(T)$*—thus every invariant subspace is hyperinvariant as well, i.e., invariant for every operator* $X = (T)'$.

(iv) *For every proper invariant subspace* $\mathfrak{L}$ *for* T *the minimal function of* $T|\mathfrak{L}$ *is a proper divisor of* m_T.

(v) *If* $\mathfrak{L}_1$ *and* $\mathfrak{L}_2$ *are different invariant subspaces for* T *then* $T|\mathfrak{L}_1$ *and* $T|\mathfrak{L}_2$ *are not quasi-similar.*

(vi) *Every operator* $X \in (T)'$ *is a "function" of* T*; in fact* $X = \phi(T)$ *with some* $\phi \in N_T$.

Here $(T)'$ denotes the "commutant" of T, i.e., the family of operators commuting with T. The "second commutant" $(T)''$ consists of operators commuting with the operators X in $(T)'$.

Using Theorem A one can prove

Theorem C. *For every operator* T *of class* C_0 *and with* $\mu_T < \infty$*, the operators in* $(T)''$ *are functions* $\phi(T)$ *of* T*, with* $\phi \in N_T$.

It is not known as yet whether here the condition $\mu_T < \infty$ can be dropped.

The proof of these theorems is rather long so we cannot go into them here. Let us only mention that an important step is to prove the following:

Proposition. *For every operator* T *of class* C_0 *we have*

$$S(m) \oplus T_2 \succ T \succ S(m) \oplus T_1$$

where $m = m_T$ *and* T_1, T_2 *are operators of class* C_0.

10. Examples of Quasi-Similarity and the Class N_T of Functions

1. In order to elucidate the role of quasi-similarity in Theorems A and B of §9 we are going to consider two relatively prime inner functions a, b, and the operators

$$S(ab) \text{ and } S(\Theta), \qquad \text{where } \Theta = \begin{bmatrix} a & 0 \\ 0 & b \end{bmatrix};$$

obviously,

$$\mathfrak{H}(\Theta) = \mathfrak{H}(a) \oplus \mathfrak{H}(b) \quad \text{and} \quad S(\Theta) = S(a) \oplus S(b).$$

From the ("lifting") Theorem 3′ of §7 we deduce that the general form of operators

$$X: \mathfrak{H}(ab) \to \mathfrak{H}(\Theta), \qquad X': \mathfrak{H}(\Theta) \to \mathfrak{H}(ab)$$

satisfying the equations

$$S(\Theta)X = XS(ab), \qquad S(ab)X' = X'S(\Theta) \tag{10.1.1}$$

is the following:

$$Xw = P_{\mathfrak{H}(\Theta)}\begin{bmatrix} y_1 \\ y_2 \end{bmatrix} w, \qquad w \in \mathfrak{H}(ab), \tag{10.1.2}$$

$$X'\begin{bmatrix} u \\ v \end{bmatrix} = P_{\mathfrak{H}(ab)}[y'_1, y'_2]\begin{bmatrix} u \\ v \end{bmatrix}, \quad \begin{bmatrix} u \\ v \end{bmatrix} \in \mathfrak{H}(\Theta), \tag{10.1.3}$$

where y_1, y_2, y'_1, y'_2 are functions in H^∞ satisfying the conditions

$$\begin{bmatrix} y_1 \\ y_2 \end{bmatrix} ab = \begin{bmatrix} a & 0 \\ 0 & b \end{bmatrix}\begin{bmatrix} \alpha \\ \beta \end{bmatrix}, \qquad [y'_1, y'_2]\begin{bmatrix} a & 0 \\ 0 & b \end{bmatrix} = ab[\gamma, \delta], \tag{10.1.4}$$

with some $\alpha, \beta, \gamma, \delta \in H^\infty$. The first condition is void (set $\alpha = by_1$, $\beta = ay_2$), while the second means that

$$y'_1 = b\gamma, \quad y'_2 = a\delta \qquad \text{with some } \gamma, \delta \in H^\infty. \tag{10.1.5}$$

Suppose we can choose X, X' such that $X'X = I$, which is the case in particular if $S(\Theta)$ and $S(ab)$ are similar. Then, applying the "multiplication property" of the lifting (§7.3) we have, for any $w \in \mathfrak{H}(ab)$,

$$w = X'Xw = P_{\mathfrak{H}(ab)}[y'_1, y'_2]\begin{bmatrix} y_1 \\ y_2 \end{bmatrix} w$$

$$= (y'_1 y_1 + y'_2 y_2)w + (\text{an element of } abH^2).$$

Choosing, in particular,

$$w(\lambda) = 1 - \overline{a(0)}\,\overline{b(0)}\,a(\lambda)\,b(\lambda)$$

(cf. (9.2.2)) and using (10.1.5) we infer that $1 = a\xi + b\eta + ab\zeta$ with some $\xi, \eta \in H^\infty$ and $\zeta \in H^2$. As $\zeta = \bar{a}\bar{b} - \bar{b}\xi - \bar{a}\eta$ on the unit circle we also have $\zeta \in H^\infty$. Hence,

$$1 = ax + by \tag{10.1.6}$$

with some $x, y \in H^\infty$. Thus the existence of $x, y \in H^\infty$ satisfying the equation (10.1.6) is a *necessary* condition for $S(a) \oplus S(b)$ to be *similar* to $S(ab)$.

A case when this condition is certainly *not* fulfilled is the following:

a: a Blaschke product with real zeros λ_n converging to 1, say with $\lambda_n = 1 - 1/n^2$,
b: the "singular" inner function $\exp[(\lambda + 1)/(\lambda - 1)]$.

Indeed, in this case a and b tend to 0 on the points λ_n (as $n \to \infty$) and so does $ax + by$ for every $x, y \in H^\infty$, contradicting (10.1.6).

Thus in this case the operators $S(a) \oplus S(b)$ and $S(ab)$ are *not similar.*

But it turns out that these operators are *quasi-similar* for every relatively prime a, b. Indeed by choosing $y_1 = y_2 = 1$ and $y'_1 = b$, $y'_2 = a$ (see (10.1.5)), the operators X and X' are quasi-affinities. To prove this we argue as follows. First recall the definitions:

$$Xw = P_{\mathfrak{H}(\Theta)}\begin{bmatrix} w \\ w \end{bmatrix} = \begin{bmatrix} P_{\mathfrak{H}(a)}w \\ P_{\mathfrak{H}(b)}w \end{bmatrix} \quad \text{for } w \in \mathfrak{H}(ab),$$

$$X'\begin{bmatrix} u \\ v \end{bmatrix} = P_{\mathfrak{H}(ab)}(bu + av) \quad \text{for } \begin{bmatrix} u \\ v \end{bmatrix} \in \mathfrak{H}(\Theta), \text{ i.e., } u \in \mathfrak{H}(a),\ v \in \mathfrak{H}(b).$$

(1) $Xw = 0$ implies $w \in aH^2 \cap bH^2 = abH^2$; as $w \in \mathfrak{H}(ab)$ this is possible only if $w = 0$. Thus X has zero kernel.

(2) If $\begin{bmatrix} u \\ v \end{bmatrix} \in \mathfrak{H}(\Theta)$ is orthogonal to Xw for every $w \in \mathfrak{H}(ab)$, then $(u + v, w) = (u, P_{\mathfrak{H}(a)}w) + (v, P_{\mathfrak{H}(b)}w) = \left(\begin{bmatrix} u \\ v \end{bmatrix}, Xw\right) = 0$ for every $w \in \mathfrak{H}(ab)$, and hence $u + v \in abH^2$. On the other hand, $u \perp aH^2$ and $v \perp bH^2$ imply $u + v \perp abH^2$. Hence, $u + v = 0$. Therefore, both u and v are orthogonal to aH^2 and bH^2, and hence to $aH^2 \vee bH^2$, i.e., to H^2. Thus, $u = v = 0$: the range of X is dense in $\mathfrak{H}(\Theta)$.

(3) Suppose $X'\begin{bmatrix} u \\ v \end{bmatrix} = 0$ for a $\begin{bmatrix} u \\ v \end{bmatrix} \in \mathfrak{H}(\Theta)$; that is, suppose $bu + av \in abH^2$. As on the other hand $u \in \mathfrak{H}(a)$, $v \in \mathfrak{H}(b)$ imply $u \perp aH^2$, $v \perp bH^2$ and therefore $bu \perp abH^2$ and $av \perp abH^2$, we have $bu + av \perp abH^2$. Thus $bu + av = 0$. As a, b are prime we infer that a is a divisor of u and b is a divisor of v, and hence $u \in aH^2$, $v \in bH^2$. We conclude that $u = v = 0$: X' has zero kernel.

(4) Let $w \in \mathfrak{H}(ab)$ be orthogonal to the range of X', i.e., orthogonal to $bu + av$ for every $u \in \mathfrak{H}(a)$ and $v \in \mathfrak{H}(b)$. Then $w \perp b\mathfrak{H}(a)$; and as $w \perp baH^2$ we also have $w \perp b(\mathfrak{H}(a) \oplus aH^2)$, i.e., $w \perp bH^2$. Analogously, $w \perp aH^2$. As $aH^2 \vee bH^2 = H^2$ because a, b are prime, we get $w = 0$. Hence X' has dense range.

Summarizing, we have proved: *If a, b are any two prime inner functions then $S(a) \oplus S(b)$ is quasi-similar to $S(ab)$, but if the equation $1 = ax + by$ has no solution $x, y \in H^\infty$ then $S(a) \oplus S(b)$ is not similar to $S(ab)$.*

This example shows that with operators on infinite-dimensional Hilbert spaces quasi-similarity is a weaker, but apparently more natural, relation than similarity. In particular, Theorem A of §9 does not hold with similarity in place of quasi-similarity.

2. In Theorems B and C of §9 we were concerned with operators A which admit a representation $A = \phi(T)$, with the given operator $T \in C_0$ and with some function ϕ of class N_T. Since the operators A are bounded, it is natural to ask whether they also admit a representation $A = w(T)$ with some function $w \in H^\infty$. That this is in general *not* the case will be shown by the following example, of considerable interest in itself.

Let again a, b be two, relatively prime inner functions and set $T = S(a) \oplus S(b)$. Then $m_T = ab$. Note that $a + b$ and ab have no nonconstant inner divisors, and hence by Corollary 3 in §8.3 the operator $(a + b)(T)$ is a quasi-affinity. Thus the function $\phi = (a - b)/(a + b)$ belongs to the class N_T. The corresponding operator $V = \phi(T)$ turns out to be bounded, indeed a symmetry; namely $V = (-I_{\mathfrak{H}(a)}) \oplus I_{\mathfrak{H}(b)}$. For

$$
\begin{aligned}
&(a - b)(T) \cdot V - (a + b)(T) \\
&\quad = [-(a - b) - (a + b)](S(a)) \oplus [(a - b) - (a + b)](S(b)) \\
&\quad = -2[a(S(a)) \oplus b(S(b))] = 0 \oplus 0 = 0.
\end{aligned}
$$

But, in general, V cannot be represented in the form $V = w(T)$, with $w \in H^\infty$. For if it *can*, then we have $-I_{\mathfrak{H}(a)} = w(S(a))$ and $I_{\mathfrak{H}(b)} = w(S(b))$; considering in particular the functions $1 - \overline{a(0)}a \in \mathfrak{H}(a)$ and $1 - \overline{b(0)}b \in \mathfrak{H}(b)$ we infer that

$$
\begin{aligned}
-1 + \overline{a(0)}a - w \cdot (1 - \overline{a(0)}a) &\in aH^2, \\
1 - \overline{b(0)}b - w \cdot (1 - \overline{b(0)}b) &\in bH^2,
\end{aligned}
$$

and hence

$$
1 + w = au, \qquad 1 - w = bv,
$$

where $u, v \in H^2$; indeed $u, v \in H^\infty$. Thus we obtain that a *necessary condition for* V *to have a representation* $V = w(T)$ *with some* $w \in H^\infty$ *is that the equation* $1 = ax + by$ *admit a solution* $x, y \in H^\infty$.

As we know, there are relatively prime a, b, for which this equation has no such solution.

This example indicates that our functional calculus with cnu contractions T, constructed in §8 for the class N_T, is natural enough; indeed the class of bounded analytic functions would not suffice.

References

The general reference is the monograph [H] quoted in the Preface, where detailed references and historical comments can also be found.

Section 1. Chapter I in [H]. The first paper on the (strong) unitary dilation is [9], where a completely different proof was given. The construction given in the text is essentially that of [8].

Section 2. Chapter II in [H]. See also [5].

Section 3. Chapter VI in [H].

Section 4. Chapter VI and Chapter IX.1 in [H].

Section 5. Chapter VII in [H]. For "strange" factorizations, see [3].

Section 6. Chapter I in [H]. Further references: [1], [2], [6], [15].

Section 7. Chapters II.2 and VI.8 in [H]. Investigations in this direction were started by the paper [7]. There are many applications of these theorems; see e.g. [14].

Section 8. Chapters III and IV in [H].

Section 9. Chapters III, VIII, and IX in [H] and [10]–[14].

Section 10. See [11] and [12].

1. T. Ando, *On a pair of commutative contractions*, Acta Sci. Math. (Szeged) 24 (1963), 88–90. MR 27 #5132.

2. S. Brehmer, *Über vertauschbare Kontraktionen des Hilbertschen Raumes*, Acta Sci. Math. (Szeged) 22 (1961), 106–111. MR 24 #A1023a.

3. C. Foiaş, *Factorisations étranges*, Acta Sci. Math. (Szeged) 34 (1973), 85–89.

4. I. C. Gohberg and M. G. Kreĭn, *On a description of contraction operators similar to unitary ones*, Funkcional. Anal. i Priložen. 1 (1967), 38–60 = Functional Anal. Appl. 1 (1967), 33–52. MR 35 #4761.

5. I. Halperin, *The unitary dilation of a contraction operator*, Duke Math. J. 28 (1961), 563–571. MR 24 #A2853.

6. J. von Neumann, *Eine Spektraltheorie für allgemeine Operatoren eines unitären Raumes*, Math. Nachr. 4 (1951), 258–281. MR 13, 254.

7. D. Sarason, *Generalized interpolation in* H^∞, Trans. Amer. Math. Soc. 127 (1967), 179–203. MR 34 #8193.

8. J. J. Schäffer, *On unitary dilations of contractions*, Proc. Amer. Math. Soc. 6 (1955), 322. MR 16, 934.

9. B. Sz.-Nagy, *Sur les contractions de l'espace de Hilbert*, Acta Sci. Math. (Szeged) 15 (1953), 87–92. MR 15, 326.

10. ———, *Hilbertraum-Operatoren der Klasse* C_0, Abstract Spaces and Approximation (Proc. Conf., Oberwolfach, 1968), Birkhäuser, Basel, 1969, pp. 72–81. MR 41 #4287.

11. B. Sz.-Nagy and C. Foiaş, *Opérateurs sans multiplicité*, Acta Sci. Math. (Szeged) 30 (1969), 1–18.

12. ———, *Modèle de Jordan pour une classe d'opérateurs de l'espace de Hilbert*, Acta Sci. Math. (Szeged) 31 (1970), 91–115. MR 41 #9038.

13. ———, *Compléments à l'étude des opérateurs de classe* C_0, Acta Sci. Math. (Szeged) 31 (1970), 287–296 and 33 (1972), 113–116. MR 44 #845.

14. ———, *On the structure of intertwining operators*, Acta Sci. Math. (Szeged) 35 (1973), 225–254.

15. N. Th. Varopoulos, *Sur une inégalité de von Neumann*, C. R. Acad. Aci. Paris Ser. A–B 277 (1973), A19–A22.